U.S. Department of Transportation
National Highway Traffic Safety Administration

DOT HS 811 546

July 2012

Development of the WinSMASH 2010 Crash Reconstruction Code

This publication is distributed by the U.S. Department of Transportation, National Highway Traffic Safety Administration, in the interest of information exchange. The opinions, findings and conclusions expressed in this publication are those of the author(s) and not necessarily those of the Department of Transportation or the National Highway Traffic Safety Administration. The United States Government assumes no liability for its content or use thereof. If trade or manufacturers' names or products are mentioned, it is because they are considered essential to the object of the publication and should not be construed as an endorsement. The United States Government does not endorse products or manufacturers.

REPORT DOCUMENTATION PAGE		Form Approved OMB No. 0704-0188
Public reporting burden for this collection of information is estimated to average 1 hour per response, including the time for reviewing instructions, searching existing data sources, gathering and maintaining the data needed, and completing and reviewing the collection of information. Send comments regarding this burden estimate or any other aspect of this collection of information, including suggestions for reducing this burden, to Washington Headquarters Services, Directorate for Information Operations and Reports, 1215 Jefferson Davis Highway, Suite 1204, Arlington, VA 22202-4302, and to the Office of Management and Budget, Paperwork Reduction Project (0704-0188), Washington, DC 20503.		
1. AGENCY USE ONLY (Leave blank) DOT HS 811 546	2. REPORT DATE July 2012	3. REPORT TYPE AND DATES COVERED Final Report April 2006 – January 2011
4. TITLE AND SUBTITLE Development of the WinSMASH 2010 Crash Reconstruction Code		5. FUNDING NUMBERS
6. AUTHOR(S) Hampton C. Gabler, Carolyn Hampton, and Nicholas Johnson		
7. PERFORMING ORGANIZATION NAME(S) AND ADDRESS(ES) Virginia Polytechnic Institute and State University Department of Mechanical Engineering Blacksburg, VA 24061		8. PERFORMING ORGANIZATION REPORT NUMBER
9. SPONSORING/MONITORING AGENCY NAME(S) AND ADDRESS(ES) U.S. Department of Transportation National Highway Traffic Safety Administration 1200 New Jersey Avenue SE. Washington, DC 20590		10. SPONSORING/MONITORING AGENCY REPORT NUMBER
11. SUPPLEMENTARY NOTES Contracting Officer's Technical Representative – Dinesh Sharma		
12a. DISTRIBUTION/AVAILABILITY STATEMENT Document is available through the National Technical Information Service www.ntis.gov.		12b. DISTRIBUTION CODE

13. ABSTRACT (Maximum 200 words)

This report describes the development of WinSMASH2010, an extensive update and enhancement to the WinSMASH crash reconstruction code. The specific objectives were (1) to correct known programming bugs in the original WinSMASH, (2) convert the code from the obsolete Delphi language to C-Sharp to allow future upgrades, and (3) to enhance WinSMASH accuracy by implementing an automated method of selecting vehicle specific stiffness coefficients.

14. SUBJECT TERMS Crash reconstruction, WinSMASH, ΔV			15. NUMBER OF PAGES 68
			16. PRICE CODE
17. SECURITY CLASSIFICATION OF REPORT Unclassified	18. SECURITY CLASSIFICATION OF THIS PAGE Unclassified	19. SECURITY CLASSIFICATION OF ABSTRACT Unclassified	20. LIMITATION OF ABSTRACT

NSN 7540-01-280-5500

Standard Form 298 (Rev. 2-89)
Prescribed by ANSI Std. 239-18
298-102

Acknowledgments

The authors wish to acknowledge Brian Marple, Computer Science Corporation, for his assistance in the NASSMAIN-WinSMASH integration.

Table of Contents

Acknowledgments ii

List of Figures.. iv

List of Tables v

1. Introduction and Background... 1

2. Description of Revisions to WinSMASH 4

3. Implementation of Improved Vehicle Stiffness Selection Procedure 5

4. Generation of WinSMASH Vehicle Stiffness Library 9

5. Changes to the NASS/CDS ΔV Distributions Using the WinSMASH 2008 Algorithm 18

6. AutoSMASH 35

7. Summary 39

8. References... 40

Appendix A – Revisions to WinSMASH 41

Appendix B– AutoSMASH Input/Output Table Format.. 48

Appendix C– AutoSMASH Programmer's Guide.. 51

Appendix D - WinSMASH Data Tables 58

Appendix E – WinSMASH Research Studies 59

List of Figures

Figure 1. WinSMASH 2007 screen for manual entry of the stiffness category number 6

Figure 2. WinSMASH 2010 screen for automated selection of stiffness category. Stiffness-related fields are circled. 6

Figure 3. WinSMASH ΔV versus EDR ΔV 24

Figure 4. Changes in average ΔV due to WinSMASH enhancements 24

Figure 5. WinSMASH ΔV versus EDR ΔV for cars 25

Figure 6. WinSMASH ΔV versus EDR ΔV pickup trucks 25

Figure 7. WinSMASH ΔV versus EDR ΔV for utility vehicles 25

Figure 8. WinSMASH ΔV versus EDR ΔV for vans 25

Figure 9. WinSMASH ΔV versus EDR ΔV for vehicles with partial overlap 27

Figure 10. WinSMASH ΔV versus EDR ΔV for vehicles with major overlap 27

Figure 11. WinSMASH ΔV versus EDR ΔV for low-confidence reconstructions 28

Figure 12. WinSMASH ΔV versus EDR ΔV for high-confidence reconstructions 28

Figure 13. WinSMASH ΔV versus EDR ΔV for vehicles with lower EDR-reported ΔVs 29

Figure 14. WinSMASH ΔV versus EDR ΔV for vehicles with higher EDR-reported ΔVs 29

Figure 15. WinSMASH ΔV versus EDR ΔV when using vehicle-specific stiffness 30

Figure 16. WinSMASH ΔV versus EDR ΔV when using categorical stiffness 30

Figure 17. Diagram of the function of AutoSMASH. 35

Figure 18. Initial screen of AutoSMASH. 36

Figure 19. The main form of AutoSMASH. 37

List of Tables

Table 1. WinSMASH 2007 categorical stiffness values. 7

Table 2. WinSMASH 2009/2010 categorical stiffness values. 8

Table 3. Table "Stiffness2010". 13

Table 4. Column formats and units in "Stiffness2010" table. 14

Table 5. Table "Stiffness2007." 14

Table 6. Column formats in "Stiffness2007" table. 15

Table 7. Table "Size2010." 16

Table 8. Column formats in "Size2010" table. 17

Table 9. Composition of the Dataset. 23

Table 10. ΔV by body type. 25

Table 11. ΔV by calculation type. 26

Table 12. Regressions for low-/high-ΔV crashes. 29

Table 13. Column names, data types and units of all fields required for manually supplying cases to AutoSMASH via Microsoft Access or Microsoft Excel. Columns that indicate "output" in the description appear only in saved results, and are not required in input tables. 48

Table 14. Numerical codes for WinSMASH calculation types. 50

Table 15. AutoSMASH Default versions of WinSMASH 51

Table 16. The Oracle tables queried by AutoSMASH, and the relevant fields. 52

Table 17. VEHICLECRASH field name changes. 54

Table 18. WinSMASH defaults for unknown fields. 55

Table 19. Calculation type is assigned based on the value of PickDeltaVBasis for each vehicle. Asterisks (*) indicate a calculation type defined by default to an ambiguous combination of PickDeltaVBasis values. 56

Table 20. Table of weights to be used for known occupants with unknown weight. 58

1. Introduction and Background

The most commonly used measure of crash severity in National Highway Traffic Safety Administration in-depth crash databases is the change in velocity, or ΔV (pronounced "deltaV"), of vehicles involved in collisions. ΔV is defined as the change in velocity of the crashed vehicle during the collision phase. WinSMASH (Microsoft Windows version of Simulating Motor Vehicle Accident Speeds on the Highway) is the computer code used by NHTSA to reconstruct the ΔV of the vehicles involved in crashes that are documented in the National Automotive Sampling System Crashworthiness Data System (NASS/CDS), Special Crash Investigations (SCI), and Crash Injury Research Engineering Network (CIREN) in-depth crash investigations. The WinSMASH code has two separate and independent methods for estimating ΔV: one based on vehicle damage, the other based on post-collision vehicle trajectory. The damage-based algorithm estimates ΔV based on post-crash deformation measurements obtained by crash investigators in combination with vehicle stiffness measurements computed from staged crash tests. The trajectory-based algorithm estimates ΔV using momentum conservation principles and information about the path taken by the vehicles from the impact position to the rest position. Such information is obtained by crash investigators from the scene of the crash.

WinSMASH is the latest version of the crash reconstruction program CRASH (Calspan Reconstruction of Accident Speeds on Highway) that was originally developed by Calspan for NHTSA. The CRASH program has evolved through several generations at NHTSA, i.e., CRASH, CRASH2, CRASH3, CRASH3PC, SMASH, and finally WinSMASH. CRASH, CRASH2, and CRASH3 were originally developed for the mainframe and mini-computers of the 1970s and 1980s. CRASH3PC, SMASH, and WinSMASH were developed for personal computers (PCs). In addition to the implementation of a graphical interface for easier user interaction with the program, WinSMASH contains several damage algorithm changes made to provide a better description of impact damage (the trajectory algorithm, however, remained unchanged throughout all versions).

Two versions of the WinSMASH program are used for crash reconstruction at NHTSA – the integrated version and stand-alone version. These two versions are essentially the same from a computational perspective. The primary difference is that the integrated version, integrated with the Crashworthiness Data System (CDS) data entry system NASSMAIN, does not allow users to make any changes to most of the input fields. The WinSMASH program is invoked from within NASSMAIN. NASSMAIN directly passes the input parameters to the integrated WinSMASH that must be input manually to the standalone WinSMASH. After performing calculations, the integrated version of WinSMASH returns output parameters to NASSMAIN that can then be stored in the CDS database, if directed. NHTSA uses the stand-alone version for internal research purposes.

The Need for WinSMASH Update

WinSMASH v.2.42, the version of WinSMASH in production at NHTSA at the start of this project, required an upgrade to correct a list of programming bugs and implement a novel strategy to improve the vehicle stiffnesses used in the ΔV computation. WinSMASH v2.42 used generic vehicle size and stiffness categories to calculate the ΔV. With the introduction of newer models, the vehicle fleet has changed since these generic values were generated in 1995. These values needed to be updated to represent the current fleet of vehicles on the road. When the Windows version of the program was created and the programming language was transformed from FORTRAN to C to Visual Basic to Delphi, some bugs remained in the program. WinSMASH v.2.42 was written in Delphi, an older programming language no longer supported by Borland. NHTSA decided that the program needed to be revised and upgraded to remove known programming bugs, update the stiffness data, and port the code to a more viable programming language.

It should also be noted that the FORTRAN and Visual Basic versions of the program were intended as stand-alone executables but, when the program was converted to Delphi, the intent was to create a version that would function as an integrated part of NASSMAIN. This integrated version subsequently became the main focus of all work on the program. Stand-alone executable versions of the integrated program were still created for internal research, but these were a byproduct of the work on the integrated version. Thus, the primary objective of this project was to improve the integrated version of WinSMASH.

Objective

The goal of this project is to update and enhance both the integrated and stand-alone versions of the WinSMASH crash reconstruction code. The specific objectives were (1) to correct known programming bugs in the original WinSMASH and convert the code from the obsolete Delphi language to C-Sharp (C#) to allow future upgrades, and (2) to enhance WinSMASH accuracy by implementing an automated method of selecting vehicle specific stiffness coefficients.

Approach

In 2006, NHTSA contracted with Virginia Tech to develop a new version of WinSMASH. Part of the motivation for this project was to update WinSMASH to reflect the changing vehicle fleet. At the beginning of the project, the current version of WinSMASH used by NASS investigators was WinSMASH 2.42. NHTSA provided Virginia Tech with the source code for WinSMASH 2.44 – a follow up version to WinSMASH 2.42 that had not yet been implemented in NASS.

The plan was to completely restructure WinSMASH in two phases. The first phase sought to make near-term improvements such as correcting known programming bugs while planning was underway for a complete restructuring and re-development of the

WinSMASH code. The second phase would undertake the complete restructuring and rewrite of the code in C# and build a comprehensive WinSMASH library suitable for the current vehicle fleet.

WinSMASH 2007, the code developed under the first phase, was implemented in NASS in January 2007. WinSMASH 2008, the code developed under the second phase, was implemented in NASS in January 2008. Both versions of WinSMASH were extensively tested prior to implementation. The research team extensively validated the code using a suite of cases provided by NHTSA. Additional validation was performed by the NHTSA Contracting Officer's Technical Representative (COTR) and members of NHTSA Special Crash Investigation teams.

2. Description of Revisions to WinSMASH

At the beginning of the project, the current version of WinSMASH used by NASS investigators was WinSMASH 2.42. NHTSA provided the research team with the source code for WinSMASH 2.44 – a follow up version to WinSMASH 2.42 that had not yet been implemented in NASS. After correcting numerous programming bugs in WinSMASH 2.44, WinSMASH 2007 was implemented in January 2007. WinSMASH 2008, a complete rewrite of WinSMASH using C-Sharp with a new stiffness library, was implemented in January 2008. WinSMASH 2010 was first distributed in January 2010, and contained a large number of improvements over WinSMASH 2008, as well as fixes for bugs discovered in the two years that WinSMASH 2008 was in service. An update to WinSMASH 2010 (v.2010.6.2.00) was distributed in June 2010, and addressed several issues that had been uncovered with WinSMASH 2010's new features. WinSMASH 2011.1.1.01 implements a handful of additional refinements and small bugfixes, but does not include any changes to the core algorithms or the stiffness parameters.

Appendix A gives a comprehensive list of the changes made to WinSMASH since version 2.44. Changes are divided into four categories: bugfixes, new features, refinements and distribution improvements. Bugfixes are corrections of programming errors that were discovered in version 2.44 and corrected in 2011.1.1.01. New features encompass functionality that is present in version 2011.1.1.01 but not in version 2.44. Refinements include changes to WinSMASH that improve some aspect of the program but do not add new capabilities. Last, distribution improvements are differences in or improvements to the means by which WinSMASH is installed and uninstalled to the computers used by NHTSA's crash investigators.

3. Implementation of Improved Vehicle Stiffness Selection Procedure

Introduction

WinSMASH calculates ΔV values based on post-crash vehicle deformation and stiffness values (Prasad, 1990; Prasad, 1991; NHTSA, 1986; Sharma, 2007). While the crush values can be easily measured by investigators, the vehicle stiffness is more difficult to determine. In earlier versions of WinSMASH the vehicle fleet was divided into categories by vehicle bodystyle and wheelbase (for cars). For each category, WinSMASH provided average stiffness values intended to represent the entire group of vehicles. The crush and stiffness values were used to calculate the energy absorbed by each vehicle, and the total absorbed energy between two vehicles (or the vehicle and fixed barrier) was used to estimate the ΔV of each vehicle.

WinSMASH 2008 was implemented in January 2008 with updated stiffness values and stiffness selection procedures. The WinSMASH 2008 implementation also included a new library of vehicle-specific stiffness values for passenger vehicles of model years 1981-2007. The categorical stiffness coefficients were also changed to use NHTSA-supplied values that better reflected the current fleet. These alterations were subsequently retained in WinSMASH 2010. These new stiffness values resulted in changes to the WinSMASH ΔV predictions in WinSMASH 2008 and 2010.

In addition to the changes made to the stiffness values, the stiffness selection process of WinSMASH 2008 was reformulated. The previous method of stiffness assignment used by WinSMASH 2007 and earlier required that the crash investigator assign a numerical stiffness category to each vehicle based on their understanding of the vehicle bodystyle and wheelbase. One of the difficulties associated with this method was that the investigator was required to memorize (or have on hand) the exact range of wheelbases for each category, as well as know the number associated with that range.

A new, automated stiffness selection process was added in WinSMASH 2008 to reduce some of the difficulties associated with WinSMASH 2007 (Sharma et al., 2007), a feature that was again retained in WinSMASH 2010. Rather than rely on the assignment of a numerical stiffness number by the crash investigator, WinSMASH 2010 automatically assigns stiffness values based on the vehicle damage side, wheelbase, and bodystyle. WinSMASH 2010 first attempts to retrieve vehicle-specific stiffness values from the new library of vehicle stiffness values. The database is searched using the vehicle model year, make, model, and bodystyle. If a stiffness corresponding to these four parameters exactly is not found, WinSMASH 2010 will then automatically select the appropriate stiffness category for the vehicle bodystyle and wheelbase. The user may also bypass the automated stiffness selection procedure entirely by activating the "Advanced Stiffness" mode. This allows manual entry of stiffness coefficients, selection of stiffness data for an arbitrary vehicle or manual selection of a stiffness category. A comparison of the stiffness screens in the two versions of WinSMASH is shown in Figure 1 and Figure 2.

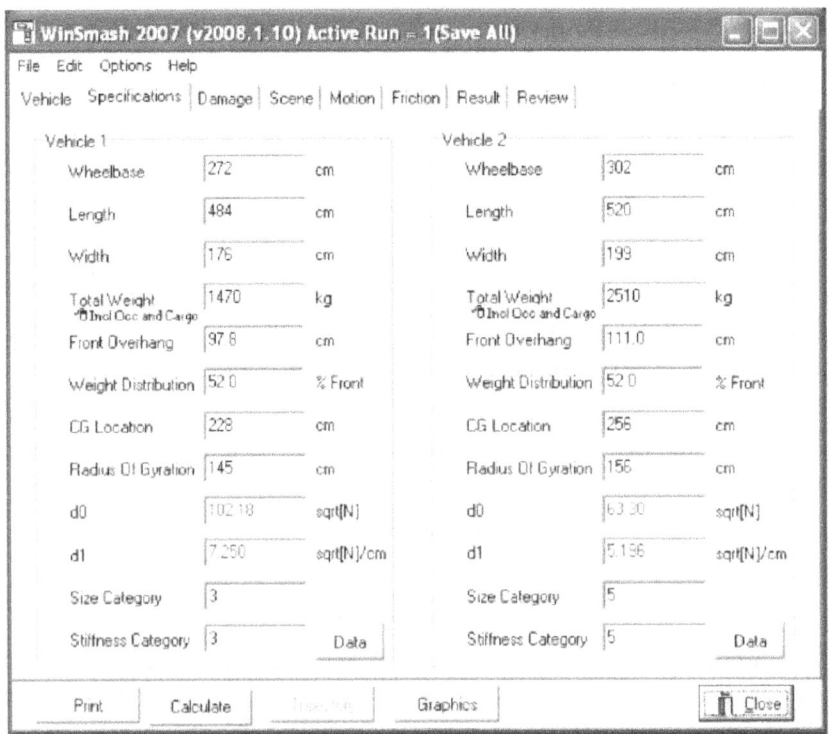

Figure 1. WinSMASH 2007 screen for manual entry of the stiffness category number

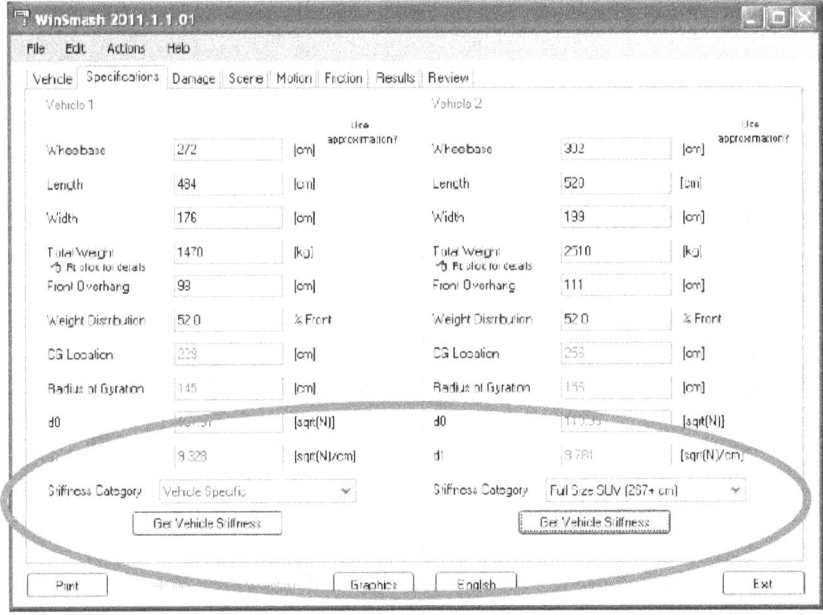

Figure 2. WinSMASH 2010 screen for automated selection of stiffness category. Stiffness-related fields are circled.

Stiffness Categories

Both the vehicle stiffness categories and the categorical stiffness values were updated first in WinSMASH 2008, then again in WinSMASH 2009. The WinSMASH 2009 stiffnesses are currently used in WinSMASH 2010. The categorical stiffness values used in WinSMASH 2007 and WinSMASH 2010 are presented in Table 1 and Table 2 respectively. In WinSMASH 2008, the old stiffness categories 7 and 8 were replaced with six new categories that are shown in Table 2. In addition, stiffness category 9 (front-wheel drive cars) was eliminated. As virtually all cars are now front-wheel drive cars, the passenger car fleet was instead assigned stiffness values according to car size category.

Table 1. WinSMASH 2007 categorical stiffness values.

Category name (wheelbase range (cm))	Front		Rear		Side	
	d0 ($\sqrt{N}$)	d1 ($\sqrt{N}/cm$)	d0 ($\sqrt{N}$)	d1 ($\sqrt{N}/cm$)	d0 ($\sqrt{N}$)	d1 ($\sqrt{N}/cm$)
1: Minicar (0 – 240)	91.46537	6.79628	93.88106	5.433098	63.3	6.836201
2: Subcompact Car (241 – 258)	97.03749	7.225553	96.2347	5.285418	63.3	8.021931
3: Compact Car (259 – 280)	102.1758	7.250295	99.48828	5.564199	63.3	7.502554
4: Intermediate Car (281 – 298)	107.0321	6.362827	99.9994	5.375778	63.3	7.216195
5: Full Size Car (299 – 313)	109.6652	6.186434	99.97363	4.507317	63.3	5.196146
6: Large Car (313 and larger)	116.0334	5.752686	74.868	6.9471	54.452	5.6925
7: Vans and SUVs[1]	109.7424	8.511337	98.68649	7.793155		
8: Pickups[1]	105.6987	7.982362	101.4218	7.773829		
9: Front-Drive Car[2]	99.18291	6.469472				

Notes:
1. Vans, SUVs, and pickup categories (7-8) may only be used when the vehicle damage plane is frontal or rear. For the right or left damage planes, the appropriate car stiffness category (1-6) must be selected based on the wheelbase of the vehicle.

2. The front-drive car category (category 9) has not been used since 2006 and prior to that was only permitted if the vehicle damage plane was frontal. For other damage planes, the appropriate car stiffness category (1-6) was selected based on the wheelbase of the vehicle.

Table 2. WinSMASH 2009/2010 categorical stiffness values.

Category name (wheelbase range)	Front		Rear		Side	
	d0 ($\sqrt{N}$)	d1 ($\sqrt{N}/cm$)	d0 ($\sqrt{N}$)	d1 ($\sqrt{N}/cm$)	d0 ($\sqrt{N}$)	d1 ($\sqrt{N}/cm$)
Minicar (0 – 240)	90.26777	6.797558	92.65763	5.561781	63.3	7.099037
Subcompact Car (241 – 258)	95.8874	7.228292	95.05376	5.952108	63.3	8.084878
Compact Car (259 – 280)	101.2924	8.149115	96.36251	5.822748	63.3	8.435508
Intermediate Car (281 – 298)	106.4641	7.555745	98.42569	5.268541	63.3	8.366999
Full Size Car (299 – 313)	109.2402	6.763367	95.02792	6.046448	63.3	8.013979
Large Car (314 and larger)	115.2161	6.733617	95.02792	6.046448	63.3	8.013979
Minivan (0 – 315)	110.0946	9.184065	98.64772	9.08219	63.3	8.982191
Full Size Van (316 and larger)	117.2993	9.888826	98.64772	9.08219	63.3	8.982191
Small SUV (0 – 266)	114.7126	9.607261	100.4481	8.449142	63.3	10.49133
Full Size SUV (267 and larger)	118.3281	9.780724	102.7815	9.813742	63.3	11.0263
Small Pickup (0 – 289)	103.9148	7.884083	99.11972	6.534046	63.3	7.961015
Full Size Pickup (290 and larger)	112.8997	8.054416	98.6221	7.165594	63.3	7.961015

4. Generation of WinSMASH Vehicle Stiffness Library

WinSMASH relies on vehicle stiffness information obtained from the WinSMASH stiffness library in order to run the necessary calculations for each case. This chapter describes the contents of the WinSMASH library, and the procedure for generating the stiffness library.

Contents of the WinSMASH Library

The WinSMASH vehicle stiffness library is contained within a single Microsoft Access database file (.mdb), and is composed of a number of tables containing different information. The tables in the WinSMASH library are:

- **AllMakeModels**: contains all distinct make and model combinations. This is used to populate the comboboxes, and does not necessarily have any corresponding matches in pVehicle.

- **pVehicle**: contains all vehicle-specific stiffness information obtained from NHTSA crash tests. It also contains both the NASS/CDS MakeID and ModelID codes and the Oracle MakeID and ModelID codes.

- **Stiffness2010**: contains categorical stiffnesses, and wheelbase ranges and bodystyle codes to which the categories apply. Categorical stiffness coefficients are used when a vehicle specific stiffness is not available for a given vehicle.

- **Size2010**: contains the size category definitions used by WinSMASH, as well as the generic vehicle specifications that correspond to those categories.

- **Stiffness2007**: currently unused, this table contains the categorical stiffnesses used by WinSMASH 2007. This is currently maintained to support possible backward-compatibility features that may be added in the future.

- **version**: this table contains only a single entry recording the date on which the library was compiled.

Procedure for Assembling the WinSMASH Stiffness Library

The stiffness library is compiled from raw crash test data into a format useful to WinSMASH. Information is drawn from the NHTSA crash test database, a listing of equivalent vehicles composed by Anderson (2009), the NASS/CDS vehicle lookup tables and tables of vehicle stiffnesses calculated for individual NHTSA crash tests (which are provided by NHTSA). The process is divided into five general steps, the details of which follow.

1) Prepare Crash Test Table: The generation of the WinSMASH library begins with spreadsheets containing frontal, side and rear crash tests, specifications of the

involved vehicles and the vehicle stiffnesses calculated from the test data. This spreadsheet is developed from data extracted from the vehicle crash test database. The first step is to make this data compatible with NASS/CDS coding and perform a thorough check for coding errors. Erroneously listed MAKEDs and MODELDs are corrected, tests involving models unknown to NASS are removed, tests with unknown bodystyles are removed and tests listed with erroneous bodystyles are corrected. These corrected tables are then cross-checked against the crash test database, and any tests that do not appear in the crash test database are removed. Finally, duplicate test numbers are removed, and the cleaned lists of crash tests are set aside.

2) <u>Generate Crash Test to NASS/CDS Cross-Reference Table</u>: In step 2, a translation table matching vehicles in the crash test database to the same vehicles in NASS/CDS is assembled for use in subsequent steps. The vehicle crash test database, the SAS copies of the NASS/CDS tables and the electronically-stored Oracle NASS/CDS tables all use different ID codes for vehicle makes and models. WinSMASH exchanges data with the Oracle database (via NASSMAIN), but the vehicle stiffness information in the stiffness library comes from the crash test database. Thus, in order for the stiffness library to be of use to WinSMASH, it is necessary to map Oracle IDs to identify vehicles instead of crash test database IDs. This table maps crash test database IDs to NASSMAIN/Oracle IDs and make/model names. It is also used to make the AllMakeModels table, which contains only the make/model names and the corresponding Oracle ID codes.

3) <u>Generate a Vehicle Clone Table</u>: The goal of step 3 is to prepare a table of vehicle clones, or a table containing vehicle models and the year ranges over which the models remained unchanged For example, the 2000 Buick Century is structurally identical to Buick Century models from years 1999 through 2005; if a test of a 2000 Buick Century were not available, a test of a 1999 Buick Century could be used instead. Thus, in the event that a certain vehicle model is not represented in the table of crash tests imported in step 1, this table of clones can be used to assign the stiffness information from a test of an equivalent vehicle. Preparation of the clone table begins using the list of vehicle clones developed/maintained by Anderson (2009). This table lists vehicles by year, make, model, and bodystyle. For each vehicle, the table lists the range of model years over which the model was cloned.

Because this list does not make use of any NHTSA make or model codes, the information in it must be matched using the make and model name strings. The first stage in step 3 is therefore to translate the make and model names to be consistent with the names used in the vehicle crash test database. Next, the clone table is checked for duplicate entries, which are then removed. It should be noted that many – but not all – of the duplicate entries are due to certain vehicle models being available in different bodystyles; these are presently ignored. After removing duplicated entries, a number of vehicles with incorrect model year ranges are corrected. Each one of these errors must be individually corrected, as there is no way to know that a model year range is incorrect based solely on information in the table. The table is then corrected for any other errors that can be found, model year ranges are verified to be consistent between all clones of a given model, vehicle years are checked for consistency with their listed model year ranges (they should lie within it), and starting years are checked to make sure that they are earlier than the end years.

4) <u>Generate a Table of Corporate Twins:</u> Step 4 also generates a table for the purpose of extrapolating the existing crash tests to other vehicles. However, in this step a table of corporate twins, rather than clones, is created. The starting material for this step is a small spreadsheet containing a number of high-volume vehicles and their corporate twins. Initially, a vehicle make, model and year range is associated with one or possibly two other make – model – year range combinations (representing a structurally equivalent vehicle). This is expanded so that every make-model-year range combination in the table appears in the primary column, making the table similar in format to the clone table developed in the previous step.

5) <u>Compile WinSMASH Library:</u> Step 5 takes the frontal, side and rear crash test information compiled in step 1, the make/model ID equivalency tables generated in step 2, and the clone table and twin table of steps 3 and 4, and from them synthesizes the "pVehicle" and "AllMakeModels" tables of the WinSMASH stiffness library. First, the make and model names in the step 1 crash test tables are updated along with some additional updates to the clone table make/model names. As described in step 3, the names are the only means by which the clone table may be matched to a crash test, so the names of all vehicles must be consistent between the two. Next, the clone table and the front, side and rear test tables are merged by year, make name and model name. This gives tables of front, side and rear crash tests, with each test also being listed with the range of clone years for the vehicle model tested. These tables are then used with the corporate twin table to assign stiffness information to vehicles that were not tested directly. Any vehicle (combination of year, make and model) appearing in the (primary column of the) twin table, but not having an exact match in the crash test tables, is matched up with a test of one of its corporate twins (if one can be found). These tables are then merged by make name and model name to the "vehdb2nass" table in the NHTSA crash test database, tying in the Oracle MAKEIDs and MODELIDs. Any tests representing vehicles that do not appear in the Oracle vehicles list are removed at this stage. WinSMASH uses a smaller set of bodystyle codes than does the crash test database, so the bodystyles listed for all the entries must be mapped to the set of values acceptable in WinSMASH as shown in Table 3. Finally, the frontal, side and rear tables are combined by year range, make, model and bodystyle to create a single entry for each vehicle containing frontal, side and rear stiffnesses. Vehicle specifications used by WinSMASH are added to each entry. Next, individual entries for every year within a given make/model/bodystyle/year range are created by copying the

existing entries. Duplicate entries are removed, and the result is the pVehicle table.

Categorical Stiffness Tables

Tables "Size2010," "Stiffness2010" and "Stiffness2007" contain size category and stiffness category definitions. By having the categories defined outside the program code, future alterations to their definitions will not require modification of WinSMASH itself, but only some small changes to the database file. These tables can contain an arbitrary number of categories (so long as it is at least 1), but the data within the tables must be properly formatted or it will not work with WinSMASH.

The "Stiffness2010" table contains the most up-to-date categorical stiffness definitions (Table 3), along with information defining the range of vehicles each category is meant to encompass. Table 3 shows the entire table as of WinSMASH 2011.1.1.01, and Table 4 gives the required column formats. The "stiffCategory" field contains the category number. The category numbers do not have to be in order, and they do not have to be sequential. Positive numbers are recommended, however. "name" is self-explanatory: it contains the category's displayed name, and can be any string. The "minWheelbase" and "maxWheelbase" fields store the range of wheelbases covered by a given category, in whole centimeters. For categories that apply to some or all of the same bodystyles, values should not overlap – notice how category 1 ends at 240 cm and category 2 begins at 241 cm. Having overlapping wheelbase ranges for some bodystyles should not cause an error, but WinSMASH will always choose the first of the two categories in the table when using its automated selection logic. For categories that can accommodate arbitrarily large wheelbases, the code "-99" is entered for "maxWheelbase." The "bodystyles" category contains a comma-delimited list of every bodystyle to which the stiffness category is intended to apply. This is only for the automated selection of a stiffness category; the WinSMASH user may of course apply any category to any vehicle, regardless of bodystyle, at their discretion. The purpose of the remaining columns is readily apparent – they store the stiffness coefficients from Table 2 associated with the category for each vehicle side.

Table 3. Table "Stiffness2010".

stiffCategory	name	minWheelbase	maxWheelbase	bodystyles	frontD0	frontD1	rearD0	rearD1	sideD0	sideD1
1	Minicar	0	240	2S,4S,3H,5H,CV,SW,LM,2C	90.26777	6.797558	92.65763	5.561781	63.3	7.099037
2	Subcompact Car	241	258	2S,4S,3H,5H,CV,SW,LM,2C	95.8874	7.228292	95.05376	5.952108	63.3	8.084878
3	Compact Car	259	280	2S,4S,3H,5H,CV,SW,LM,2C	101.2924	8.149115	96.36251	5.822748	63.3	8.435508
4	Intermediate Car	281	298	2S,4S,3H,5H,CV,SW,LM,2C	106.4641	7.555745	98.42569	5.268541	63.3	8.366999
5	Full Size Car	299	313	2S,4S,3H,5H,CV,SW,LM,2C	109.2402	6.763367	95.02792	6.046448	63.3	8.013979
6	Large Car	314	-99	2S,4S,3H,5H,CV,SW,LM,2C	115.2161	6.733617	95.02792	6.046448	63.3	8.013979
12	Minivan	0	315	MV,VN	110.0946	9.184065	98.64772	9.08219	63.3	8.982191
13	Full Size Van	316	-99	MV,VN	117.2993	9.888826	98.64772	9.08219	63.3	8.982191
14	Small SUV	0	266	UV,4U,2U	114.7126	9.607261	100.4481	8.449142	63.3	10.49133
15	Full Size SUV	267	-99	UV,4U,2U	118.3281	9.780724	102.7815	9.813742	63.3	11.0263
16	Small Pickup	0	289	PU,4P,EX	103.9148	7.884083	99.11972	6.534046	63.3	7.961015
17	Full Size Pickup	290	-99	PU,4P,EX	112.8997	8.054416	98.6221	7.165594	63.3	7.961015

Table 4. Column formats and units in "Stiffness2010" table.

Field Name	Column Format (Microsoft Access 2000)	Units
stiffCategory	Number (Long Integer)	-
name	Text	-
minWheelbase	Number (Long Integer)	cm
maxWheebase	Number (Long Integer)	cm
bodystyles	Text	-
frontD0	Number (Double)	$\sqrt{N}$
frontD1	Number (Double)	$\sqrt{N}$/cm
sideD0	Number (Double)	$\sqrt{N}$
sideD1	Number (Double)	$\sqrt{N}$/cm
rearD0	Number (Double)	$\sqrt{N}$
rearD1	Number (Double)	$\sqrt{N}$/cm

'Stiffness2007" is a legacy table that holds the original WinSMASH stiffness categories used with WinSMASH 2007 (Table 1). Table 5 shows the data and format of the table, and Table 6 lists the column data types and units. This table is currently not used, but has been retained to support possible future features related to backwards-compatibility with early versions of WinSMASH.

Table 5. Table "Stiffness2007."

Category	Field1	frontD0	frontD1	rearD0	rearD1	SideD0	SideD1
1	Minicar	91.46537	6.79628	93.88106	5.433098	63.3	6.836201
2	Subcompact Car	97.03749	7.225553	96.2347	5.285418	63.3	8.021931
3	Compact Car	102.1758	7.250295	99.48828	5.564199	63.3	7.502554
4	Intermediate Car	107.0321	6.362827	99.9994	5.375778	63.3	7.216195
5	Full Size Car	109.6652	6.186434	99.97363	4.507317	63.3	5.196146
6	Large Car	116.0334	5.752686	74.868	6.9471	54.452	5.6925
7	Vans and 4WD	109.7424	8.511337	98.68649	7.793155	0	0
8	Pickups	105.6987	7.982362	101.4218	7.773829	0	0
9	Front-Drive Vehicles	99.18291	6.469472	0	0	0	0
10	Movable Barrier	0	0	0	0	0	0
11	Immovable Barrier	0	0	0	0	0	0
12	Minivan	0	0	0	0	0	0
13	Full Size Van	0	0	0	0	0	0
14	Small SUV	0	0	0	0	0	0
15	Full Size SUV	0	0	0	0	0	0
16	Small Pickup	0	0	0	0	0	0
17	Full Size Pickup	0	0	0	0	0	0

Table 6. Column formats in "Stiffness2007" table.

Field Name	Column Format (Microsoft Access 2000)	Units
Category	Number (Long Integer)	-
Field1	Text	-
frontD0	Number (Double)	$\sqrt{N}$
frontD1	Number (Double)	$\sqrt{N}/cm$
rearD0	Number (Double)	$\sqrt{N}$
rearD1	Number (Double)	$\sqrt{N}/cm$
SideD0	Number (Double)	$\sqrt{N}$
SideD1	Number (Double)	$\sqrt{N}/cm$

'Size2010" contains the latest definitions of the vehicle size categories, and the values of all vehicle specifications that go with them. Table 7 and Table 8 show the structure and formats of this table respectively. "sizeCategory," "name," "minWheelbase" and "maxWheelbase" are handled precisely the same as the corresponding columns in "Stiffness2010." "length," "width," "percentFront" and "frontOverhang" are the vehicle total length, max width, weight distribution and front overhang – all commonly used specifications in WinSMASH. Attention should be paid to the column formats when entering this information. All lengths are in centimeters, and "percentFront" is a percentage between 0 and 100, not a decimal fraction between 0 and 1. "trackWidth," "frontCornerStiff" and "rearCornerStiff" are the average vehicle track width, front cornering stiffness and rear cornering stiffness, and are used only when a trajectory simulation is run. Being a length, "trackWidth" is in centimeters; the two cornering stiffnesses are in units of Newtons per radian (N/rad).

Table 7. Table "Size2010."

sizeCategory	name	minWheelbase	maxWheelbase	bodystyles	length	width	percentFront	frontOverhang	trackWidth	frontCornerStiff	rearCornerStiff
1	Minicar	0	240	2S,4S,3H,5H,CV,SW,LM,2C	403	161	55.3016	83.11	142	-23905	-22415
2	Subcompact Car	241	258	2S,4S,3H,5H,CV,SW,LM,2C	434	167	59.3181	90.45	146	-33362	-30831
3	Compact Car	259	280	2S,4S,3H,5H,CV,SW,LM,2C	470	174	57.3851	96.93	151	-38762	-35830
4	Intermediate Car	281	298	2S,4S,3H,5H,CV,SW,LM,2C	518	187	57.5679	106.57	157	-46413	-42885
5	Full Size Car	299	313	2S,4S,3H,5H,CV,SW,LM,2C	558	189	58.2207	101.38	158	-53218	-49077
6	Large Car	314	-99	2S,4S,3H,5H,CV,SW,LM,2C	558	189	58.2207	101.38	158	-58054	-53597
12	Minivan	0	315	MV,VN	481	188	55.2064	91.1	159	-53218	-49077
13	Full Size Van	316	-99	MV,VN	518	197	53.7923	78.52	172	-53218	-49077
14	Small SUV	0	266	UV,2U,4U	423	171	51.7274	78.59	150	-53218	-49077
15	Full Size SUV	267	-99	UV,2U,4U	480	183	56.8665	89.34	157	-53218	-49077
16	Small Pickup	0	289	PU,4P,EX	464	168	57.0814	78.4	145	-53218	-49077
17	Full Size Pickup	290	-99	PU,4P,EX	540	188	58.2221	89.59	160	-53218	-49077

Table 8. Column formats in "Size2010" table.

Field Name	Column Format (Microsoft Access 2000)	Units
sizeCategory	Number (Long Integer)	-
name	Text	-
minWheelbase	Number (Long Integer)	cm
maxWheelbase	Number (Long Integer)	cm
bodystyles	Text	-
length	Number (Long Integer)	cm
width	Number (Long Integer)	cm
percentFront	Number (Double)	%
frontOverhang	Number (Double)	cm
trackWidth	Number (Long Integer)	cm
frontCornerStiff	Number (Long Integer)	N/rad
rearCornerStiff	Number (Long Integer)	N/rad

5. Changes to the NASS/CDS ΔV Distributions Using the WinSMASH 2008 Algorithm

Introduction

ΔV, or the change in velocity of a vehicle, is a widely used indicator of crash severity. It is also popular as a predictor of occupant risk due to its correlation to occupant injuries (Gabauer & Gabler, 2008). ΔV estimates are usually obtained from crash reconstruction programs such as CRASH3 or WinSMASH. Numerous studies of CRASH3 (Smith & Noga, 1982; O'Neill et al., 1996; Lenard et al., 1998) and WinSMASH (Nolan et al., 1998; Stucki & Fessahaie, 1998) have demonstrated that these programs have substantial error in the ΔV estimates. An enhanced version of WinSMASH has been developed to address these inaccuracies.

A publicly available source of ΔVs for real-world crashes is the NASS/CDS. This database provides data from investigations of roughly 4,000 to 5,000 police-reported, tow-away crashes each year. These ΔV estimates are used by researchers to assess vehicle safety, develop vehicle test protocols, and perform costs and benefits analyses. The ΔV estimates in NASS/CDS are produced using the crash reconstruction software, WinSMASH. NASS case years 2000 to 2006 were computed with WinSMASH 2.42. NASS/CDS 2007 was computed with WinSMASH 2007. WinSMASH 2007 was computationally identical to WinSMASH 2.42. Case years 2008 onward were computed with the enhanced versions of WinSMASH, the first of which was WinSMASH 2008.

<u>Early Crash Reconstruction</u>

One of the earliest crash reconstruction programs was CRASH3. Estimates of vehicle ΔV were calculated using the crush measured from a vehicle and representative vehicle stiffness values obtained from crash tests to compute the energy absorbed by the vehicle, which was in turn used to estimate the ΔV of all vehicles in a crash (Prasad, 1990, 1991a, 1991b; NHTSA, 1986). Many modern reconstruction programs, including the WinSMASH software used for the NASS/CDS database, are descended from this program.

The vehicle stiffness values in CRASH3 and early versions of WinSMASH were represented by assigning an individual vehicle to one of nine stiffness categories. The categories:

1. Mini Cars
2. Subcompact Cars
3. Compact Cars
4. Intermediate Cars
5. Full-Size Cars
6. Large Cars
7. Vans and SUVs
8. Pickup Trucks
9. Front- Wheel-Drive Cars

The majority of vehicles fell within one of four categories: compact cars, vans and SUVs, pickup trucks, and front-wheel-drive cars.

The Appearance of Event Data Recorders

Event data recorders (EDRs) are devices installed in vehicles with the capability to record the change in vehicle velocity during a crash. Niehoff et al. (2005) showed that event data recorders provided maximum ΔV values within 6 percent of the true maximum ΔV as calculated from crash test instrumentation. The availability of EDR data for real world crashes provides an opportunity to evaluate the accuracy of ΔV reconstruction methods for conditions other than crash tests.

Using the EDRs as an objective measure of ΔV, Niehoff and Gabler (2006) examined the accuracy of WinSMASH 2.42 in predicting ΔV for real world crashes documented in years 2000 – 2003 of NASS/CDS. Their findings indicated that WinSMASH underestimated the ΔV by 23 percent on average. The degree of underprediction varied greatly by body type, i.e., car, pickup truck, van, or utility vehicle. Inclusion of vehicle restitution and the use of vehicle specific stiffness coefficients were recommended as methods to reduce the error in WinSMASH ΔV estimates. Vehicle stiffness values can be readily obtained from NHTSA crash tests and included in the WinSMASH library. These stiffness values should be updated each year to keep up with the changing characteristics of vehicle fleet. This study examined the effect of enhanced stiffness values on the accuracy of WinSMASH ΔV estimates.

Enhancement of WinSMASH

In 2006, NHTSA initiated a research effort to improve the accuracy of WinSMASH. WinSMASH 2008 was the first version of WinSMASH to include the vehicle specific stiffness approach. The implementation required the creation of a library of vehicle specific stiffness values representing stiffness data for over 5,000 vehicle years, makes, models, and body types to be included with WinSMASH to ensure ease of use and accessibility. The use of these stiffness values was prioritized over the use of categorical stiffness values. Hampton and Gabler (2009) showed that nearly 2/3 of all vehicles that are reconstructed by WinSMASH for NASS/CDS could be successfully matched with vehicle specific stiffness values.

WinSMASH 2008 will automatically select the appropriate stiffness category for reconstructions where the vehicle specific stiffness values were not available. The stiffness values corresponding to the car categories were updated to improve the accuracy of the ΔV estimates. Categories 7 to 9 were dropped. Vans and SUVs, formerly category 7, were separated into their own categories, each of which was further subdivided into large and small vehicles. Pickup trucks, formerly category 8, were similarly split into two new categories for large and small trucks.

Hampton and Gabler (2009) showed that these changes to the WinSMASH reconstruction software resulted in ΔVs 7.9 percent higher on average than the ΔVs estimated using WinSMASH 2007, which was equivalent to WinSMASH 2.42. The results were observed to vary by body type, the side of the vehicle sustaining damage, and the object struck.

After the enhancements were completed, a reevaluation of the sources of variability, such as the vehicle body type, degree of structural overlap, and investigator confidence was needed. The objective of this study was to provide this reevaluation by comparing the ΔV estimated from the enhanced version of WinSMASH to the maximum ΔVs recorded by EDRs.

Methods

Event Data Recorders

Data from 3,685 General Motors (GM) event data recorders were available for this study. A total of 245 Ford EDRs were available but were not included because too few were available for a thorough analysis. Other major automobile manufacturers such as Toyota and Chrysler include EDRs with their vehicles. However, the EDR data from these sources were not available or not readable. Therefore, the EDRs in the dataset were comprised entirely of GM data.

Not all of the General Motors EDRs recorded data for deployment level events, i.e., events of sufficient severity to trigger the air bag deployment. A total of 1,944 EDRs were removed because they did not record a deployment level event. Non-deployment events were excluded from this study because the data recorded is not "locked in" and could have been overwritten by a subsequent event. As noted by Niehoff and Gabler (2005), even if the EDR records deployment-level data, it does not always record the complete crash pulse.

One additional EDR was removed because the crash was so severe that the EDR was damaged. An additional 476 EDRs were removed because the crash pulse was not fully captured. Completeness of the crash pulse was determined by calculating the vehicle acceleration between the last two recorded ΔVs. All pulses ending with greater than 2 G of acceleration were excluded. The remaining 1,265 EDRs represented data of sufficient quality and severity to be used in this study.

All of the GM EDRs used in this study recorded ΔV data only in the longitudinal direction. Because of this, the dataset for this study was restricted to frontal impacts only.

Collection of WinSMASH Data

Obtaining WinSMASH ΔV predictions requires information about the vehicle. For events where the EDR-equipped vehicle struck another vehicle, information for the other vehicle and the orientations of both vehicles must be collected. The data needed included:

- Vehicle year, make, model, and body type;
- Dimensions of the vehicle;
- Crush profile (depth, width, location); and
- Vehicle headings and direction of force.

Fortunately, the data needed to compute WinSMASH results were readily available from years 2000 to 2008 of the NASS/CDS. As many reconstructions as possible were assembled to maximize the chance of matching a reconstruction with an EDR.

Matching EDR and WinSMASH Data

A key difference between the EDR data and NASS/CDS data was that EDRs recorded the first event of sufficient severity to trigger the air bag deployment whereas the NASS/CDS database reported ΔV for the two highest severity events. For vehicles experiencing multiple events in a crash, it can be a challenge to identify which event had been captured by the EDR and whether other events overlapped with the recorded ΔV pulse.

To ensure the correct event was isolated, the number of events associated with each vehicle was determined using the NASS/CDS database. A total of 530 EDRs were removed from the dataset because the vehicles were involved in more than one event, leaving 735 suitable crash pulses. Crashes with multiple events were permitted if the EDR-equipped vehicle experienced a single event only. An additional 124 EDRs were removed because there was no information or insufficient information to perform a WinSMASH reconstruction. This often occurred when the vehicle was involved in a crash that could not be reconstructed in WinSMASH such as a sideswipe or non-horizontal impact. An additional 112 EDRs were removed because the EDR-equipped vehicle was struck in the side. Finally, 20 EDRs were removed because the crush profiles documented in NASS/CDS were invalid, meaning that the specified Smash L (damage length) and Field L +/- D positioned the damaged area partially or completely outside the body of the vehicle. This left a total of 478 EDRs.

Computation of the WinSMASH ΔV

Regardless of which year of NASS/CDS a crash was investigated, the ΔV for each vehicle was computed using the enhanced WinSMASH. Since the EDRs in the vehicles recorded only longitudinal ΔV, all of the WinSMASH results presented in this study were the longitudinal ΔV rather than the total ΔV. ΔVs for vehicles that struck or were struck by EDR-equipped vehicles were calculated but were not included in the results, unless these vehicles also contained an EDR. There were 9 crashes in which the EDR was available for both vehicles.

Statistical Analyses

All analyses of the ΔV results were performed with the Statistical Analysis Software (SAS) version 9.2. The accuracy of the ΔV estimates were evaluated using linear regression techniques with all curves passing through the origin. Variability of the data was assessed by computing the R^2 value and the root mean square error (RMSE). Plotting of data was performed in Microsoft Excel. Note that the R^2 values were calculated by SAS and were not the same as the value that would be computed by Microsoft Excel (Eisenhauer, 2003).

Results

Composition of the Data Set

A total of 478 vehicles with both EDR data and WinSMASH reconstruction data were collected for this study. These vehicles represented crashes occurring in the years 2000 to 2008. Model years for the vehicles ranged from 1994 to 2008. Chevrolet and Pontiac vehicles represented 68 percent of the vehicles. The remaining vehicles were other GM makes such as Buick, Cadillac, GMC, Oldsmobile, and Saturn.

The make-up of the final dataset is summarized in Table 9. The dataset contained mostly cars and all of the EDR-equipped vehicles were struck in the front. The principal direction of applied force (PDOF1 in NASS/CDS), which is 0° for a perfectly frontal impact and increases clockwise around the vehicle to a maximum of 350°, indicated that most impacts were linear frontal impacts with a smaller number of angled frontal impacts.

Table 9. Composition of the Dataset.

	Total	%
All Vehicles	478	100
Body type		
Cars	354	74%
Pickup Trucks	50	10%
Utility Vehicles	50	10%
Vans	24	5%
WinSMASH Calculation Type		
Standard	273	57%
Barrier	49	10%
Missing Vehicle	136	29%
CDC Only	20	4%
WinSMASH Stiffness		
Vehicle Specific	316	66%
Categorical – Compact Car	86	18%
Categorical – Other Car	26	5%
Categorical – Minivan/Van	11	2%
Categorical – Utility Vehicle	29	6%
Categorical – Pickup Truck	10	2%
Direction of Applied Force		
290° – 310°	11	2%
320° – 340°	91	19%
350° – 10°	276	58%
20° – 40°	89	19%
50° – 70°	11	2%

The majority of WinSMASH reconstructions (67%) were standard or barrier reconstructions. Most other crashes were reconstructed with the missing vehicle algorithm (29%). The calculation type and its effects on ΔV are discussed in more detail in the calculation type section.

ΔV Estimates from WinSMASH 2008

The enhanced WinSMASH ΔV estimates for all 478 events were plotted against the event data recorder (EDR) maximum ΔV in Figure 3. The WinSMASH reconstructions underestimated the ΔV by 13.2 percent on average. This represented a substantial improvement over the previously reported 23 percent underestimation. The RMSE was 9.40 kph (5.84 mph) for the enhanced WinSMASH, whereas the RMSE was 8.08 kph (5.02 mph) for the NASS/CDS ΔVs. This increase in variability was attributed to the wider range of stiffness values obtained from the vehicle-specific stiffness approach.

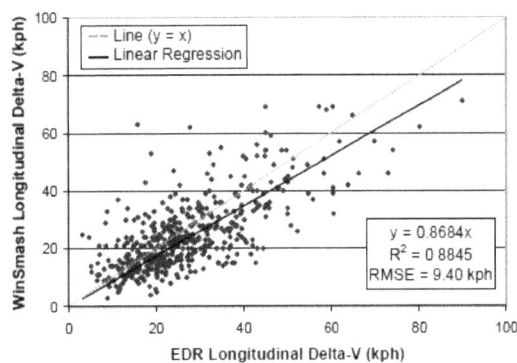
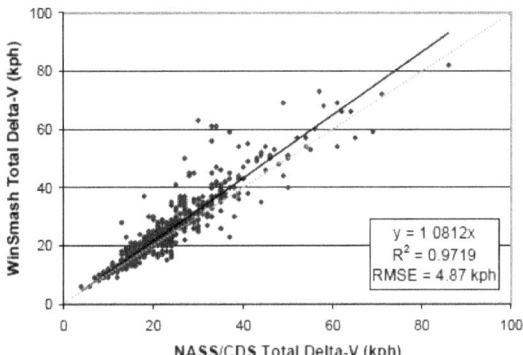

Figure 3. WinSMASH ΔV versus EDR ΔV **Figure 4. Changes in average ΔV due to WinSMASH enhancements**

The total ΔVs estimated by the enhanced WinSMASH were compared to the total ΔVs recorded in NASS/CDS. The NASS/CDS ΔVs from years 2000 to 2007 were computed with earlier versions of WinSMASH whereas the 12 percent of cases from 2008 were computed with the enhanced WinSMASH. The ΔVs from only the enhanced WinSMASH were 8.1 percent higher than the NASS/CDS ΔVs on average as shown in Figure 4. This was similar to 7.9 percent increase reported by Hampton and Gabler (2009).

Vehicle Body Type

In a study of WinSMASH 2.42 by Niehoff and Gabler (2006), the accuracy of the ΔV varied greatly by the body type of the vehicle, primarily because the body type dictated the stiffness category used. They reported that compact cars (category 3) underestimated ΔV by 14 percent, vans and utility vehicles (category 7) by 22 percent, pickup trucks (category 8) by 3 percent, and front-wheel-drive cars (category 9) by 31 percent. Since the enhanced versions of WinSMASH do not support category 9, the data from the two car categories (3 and 9) were combined into a single group representing the majority, but not all, of the cars. For this group, WinSMASH 2.42 underestimated the ΔV by 27 percent on average.

The effects of the body type on the new WinSMASH 2008 ΔV estimates are shown in Figure 5, Figure 6, Figure 7 and Figure 8, and are summarized in Table 10. In brief, the ΔV for all cars was found to be underestimated by 16.0 percent on average, 4.2 percentfor pickup trucks, and 2.3 percent for utility vehicles. Van ΔVs were underestimated by 11.2 percent on average. Variability in individual vehicle predictions remained similar across all body types except vans, for which the correlation was high due to the small number of vans available. Lumping the vans and utility vehicles together, which was roughly equivalent to category 7 in older versions of WinSMASH, resulted in a 5.2-percent ΔV error on average.

Table 10. ΔV by body type.

	Enhanced WinSMASH		WinSMASH 2.42
	% Error	RMSE (kph)	
All Vehicles	-13.2%	9.80	-23%
Cars	-16.0%	9.50	-27%
Pickup Trucks	-4.2%	9.23	-3%
Utility Vehicles	-2.3%	9.40	-22%
Vans	-11.2%	5.58	

The enhanced WinSMASH matched or improved upon the accuracy for all vehicle body types. Vans and utility vehicles showed the greatest improvement, with the underestimation reduced from -22 percent to -5 percent collectively. The average ΔV for cars also substantially increased, with the error dropping by 10 percent. Pickup truck ΔVs, which were the most accurate body type in older versions of WinSMASH, continued to be similar to the EDR ΔVs.

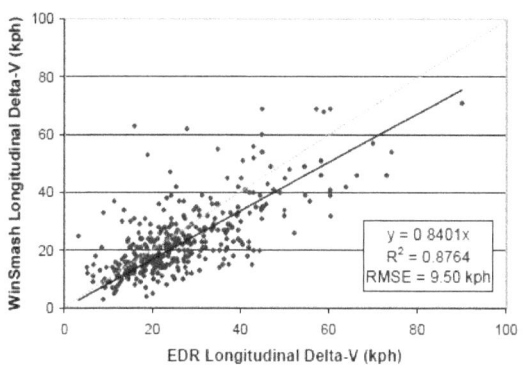

Figure 5. WinSMASH ΔV versus EDR ΔV for cars

Figure 6. WinSMASH ΔV versus EDR ΔV pickup trucks

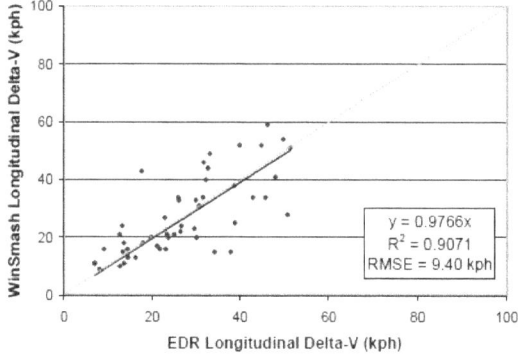
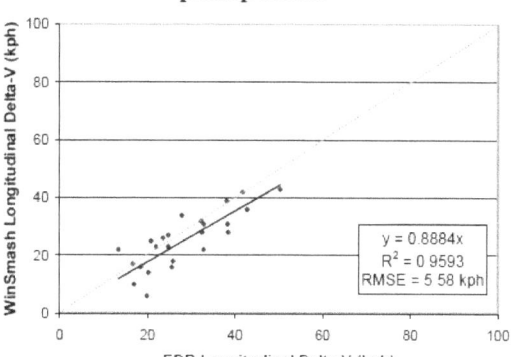

Figure 7. WinSMASH ΔV versus EDR ΔV for utility vehicles

Figure 8. WinSMASH ΔV versus EDR ΔV for vans

WinSMASH Calculation Type

In both the early and enhanced versions of WinSMASH the standard calculation type, which reconstructed vehicle-to-vehicle crashes, was the default calculation type. This calculation type required the most information, which included the crush profiles,

Collision Deformation Codes (CDCs), and the orientations and directions of applied force for both vehicles.

The standard calculation type was limited in that it was not applicable to all crashes, nor was there always sufficient information to use it for otherwise applicable crashes. The barrier calculation type allowed for crash reconstructions to be extended to vehicle-to-rigid object crashes. The CDC only and missing vehicle calculation types were available to reconstruct crashes where information about one vehicle in a vehicle-to-vehicle crash was not collected. The CDC only calculation type reconstructed crashes by approximating unknown crush profiles using the CDC code. The missing vehicle calculation type was used to perform reconstructions when information about one vehicle was unknown.

To evaluate each calculation type, a subset of 273 crashes for which all calculation types could be applied was used. The missing vehicle and CDC only calculations were each run twice: once with the EDR-equipped vehicle as a normal vehicle and a second time as the CDC only/missing vehicle. The results are summarized in Table 11.

Standard and barrier reconstructions, which require the most information about the vehicle, resulted in the best correlations with the EDR ΔVs. Both calculation types underestimated the EDR-reported ΔV by 13.3 percent and 15.4 percent respectively. Calculations performed with a missing vehicle striking the EDR-equipped vehicle were surprisingly accurate, underestimating the ΔV by only 7.4 percent while offering nearly the same correlation to the EDR ΔV. The remaining calculation types offered relatively poor correlation and were observed to overestimate the ΔV by more than 3 times in some individual crashes. The CDC only reconstructions were the only calculation type to consistently overestimate the ΔV.

Table 11. ΔV by calculation type.

Calculation Type	% Error	R^2	RMSE (kph)
Standard	-13.3%	0.92	8.12
Barrier	-15.4%	0.89	9.16
Missing (Other)	-7.4%	0.88	10.75
Missing (EDR)	-16.2%	0.82	12.29
CDC Only (Other)	+5.4%	0.82	15.35
CDC Only (EDR)	+9.5%	0.82	16.17

Extent of Structural Overlap

To determine the amount of structural overlap for each vehicle, the NASS/CDS field "Specific Longitudinal Location," or SHL1, was used. This field is part of the Collision Deformation Code (CDC) and provides a reasonable indicator of the width and location of direct damage to the vehicle (SAE, 1980). Values of C, L, and R were descriptors for damage to less than half the length of the damaged side and were classified as partial overlap. Values of Y, Z, and D applied to vehicles with damage to more than half the length of the struck side and were classified as major overlap. The SHL1 field was not available for 30 vehicles that were removed from this analysis.

Figure 9 and Figure 10 show the ΔV predictions of the enhanced WinSMASH plotted against the EDR ΔVs for vehicles with partial and major overlap. Both groups underestimated the true ΔV. However, the vehicles with full or nearly complete overlap, i.e., more than half the vehicle side sustaining direct damage, underestimated by 11.7 percent on average whereas vehicles with partial overlap or direct damage to less than half of the vehicle side underestimated by 24.1 percent on average.

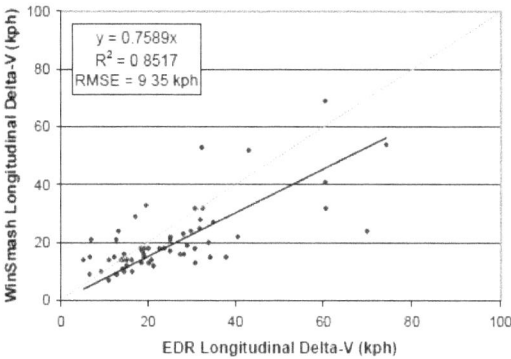
Figure 9. WinSMASH ΔV versus EDR ΔV for vehicles with partial overlap

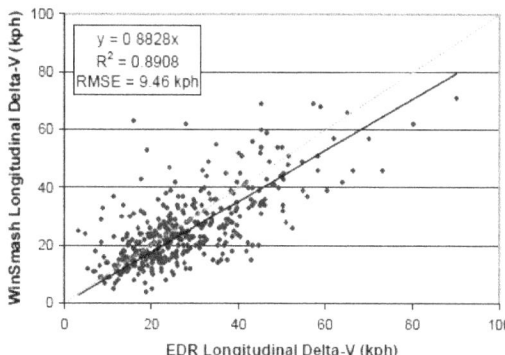
Figure 10. WinSMASH ΔV versus EDR ΔV for vehicles with major overlap

These results were consistent with the findings of Nolan et al. (1998) and Stucki and Fessahaie (1998) in their studies of previous versions of CRASH3 and WinSMASH. The disparity between the two groups, 13.4 percent for WinSMASH 2008, was greater than the difference reported in earlier studies.

Confidence in Reconstruction

When ΔVs are recorded in NASS/CDS, the investigators also record the degree of confidence in the reconstruction in the DVConfid field. The EDRs were split into two groups to determine the extent to which the ΔV errors were due to crashes with poor confidence. The high-confidence group contained all reconstructions recorded as "reasonable," whereas the other, low-confidence group contained results marked as "appears high," "appears low," or "borderline."

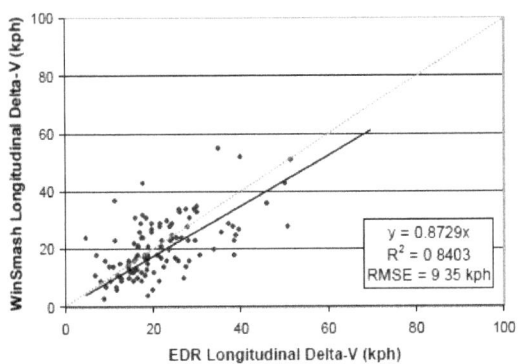 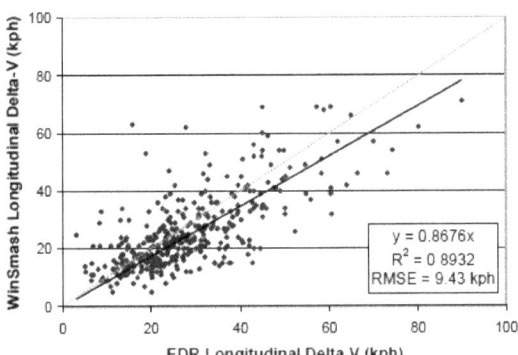

Figure 11. WinSMASH ΔV versus EDR ΔV for low-confidence reconstructions

Figure 12. WinSMASH ΔV versus EDR ΔV for high-confidence reconstructions

The results for the low- and high-confidence groups are shown in Figures 11 and 12. Both groups underestimated the ΔV by roughly 13 percent on average. The correlation between the WinSMASH and EDR ΔVs was stronger for the high-confidence group and weaker for the low-confidence group. These results agree with previous research on the effects of reconstruction confidence.

Low and High-ΔV Crashes

The stiffness values used by all versions of WinSMASH were derived from crash tests performed at speeds ranging from 48.3 to 56.3 kph (30 to 35 mph). Smith and Noga (1982) showed that the accuracy was diminished as the crash ΔV deviated from the test conditions. The error inherent in assumptions of purely plastic deformation, the shape of the stiffness curve, and the speed at which no damage occurs were most pronounced in low-speed crashes.

The EDRs were divided into a group with maximum reported ΔVs less than 24.1 kph (15 mph) and a second group consisting of EDRs with ΔVs of 24.1 kph or greater recorded. The 24.1 kph threshold was arbitrarily chosen to match the limit used in a prior study of WinSMASH 2.42. The results are shown in Figure 13 and Figure 14. The linear regression for the low-speed impacts may appear to be quite accurate, but the correlation was lower. The underestimation by the high-speed group was slightly worse than the reported error for all vehicles in Figure 3 with a better correlation.

The error of the low-ΔV group varied with the value of the ΔV threshold. Changing the threshold by as little as 3 mph (5 kph) changed the average ΔV error by as much as 20 percent. To illustrate this, the same analysis was performed with a threshold value of 16.1 kph (10 mph) and 32.2 kph (20 mph). The results are summarized in Table 12. The high-speed group remained consistent in both accuracy and correlation. For the low-speed group there was no consistency in either.

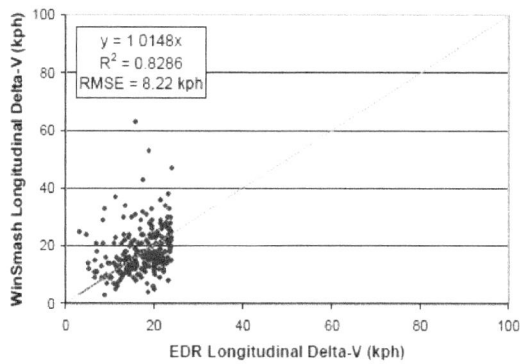

Figure 13. WinSMASH ΔV versus EDR ΔV for vehicles with lower EDR-reported ΔVs

Figure 14. WinSMASH ΔV versus EDR ΔV for vehicles with higher EDR-reported ΔVs

Table 12. Regressions for low-/high-ΔV crashes.

Threshold (kph)		Error	R2
16.1	< 16.1	+23.9%	0.768
	≥ 16.1	-14.6%	0.902
24.1	< 24.1	+1.5%	0.829
	≥ 24.1	-16.3%	0.912
32.2	< 32.2	-5.5%	0.857
	≥ 32.2	-17.7%	0.916

Categorical Versus Vehicle Specific Stiffness

Despite the large number of vehicle specific stiffness values available in the WinSMASH library, approximately a third of the vehicles could not be found and instead used categorical stiffness values. A subset of 316 vehicles, all of the EDR-equipped vehicles for which vehicle specific stiffness values were used, were recomputed with the enhanced WinSMASH using the updated categorical stiffness values so that the relative accuracies of the two stiffness methods might be assessed. The resulting ΔVs are shown in Figure 15 and Figure 16.

Both the vehicle specific and categorical approach to determining the vehicle stiffness resulted in roughly the same result: the ΔV was underestimated by 12.3 percent on average. The correlations were similar, with the categorical values appearing to be slightly more consistent. However, the differences between the ΔV predictive abilities were not significant (P=0.312 for two-tailed paired t-test). When the ΔVs were plotted against each other, there was less than 1-percent error in the linear regression.

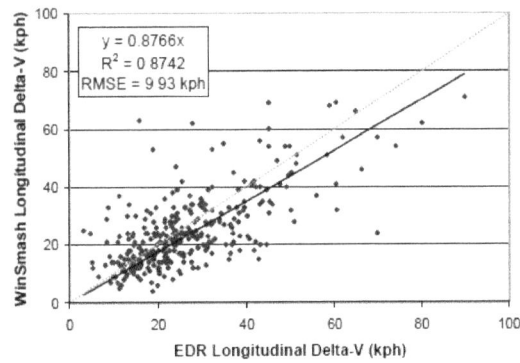
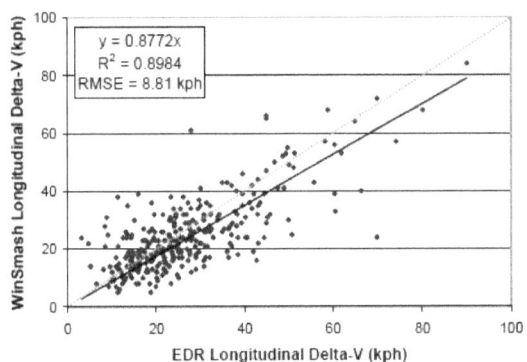

Figure 15. WinSMASH ΔV versus EDR ΔV when using vehicle-specific stiffness

Figure 16. WinSMASH ΔV versus EDR ΔV when using categorical stiffness

Discussion

Categorical Versus Vehicle Specific Stiffness

The new vehicle specific stiffness approach and updated categorical stiffness values in the enhanced WinSMASH resulted in a 9.8-percent reduction in error for average ΔV of all vehicles compared to previous versions of WinSMASH. Although the ΔV was still underestimated by 13.2 percent as compared to the ΔV from EDRs, this was still a substantial improvement. This was slightly higher than the ΔV increase observed by Hampton and Gabler (2009) and was due to the dataset used in this study being restricted to only frontal crashes.

Niehoff and Gabler (2006) showed that the error in WinSMASH 2.42 ΔV estimates was strongly dependent on the vehicle body type, which dictated which stiffness category was used in WinSMASH 2.42. The new stiffness selection process in WinSMASH 2008 greatly reduced the error for many of the vehicle body types by raising the average stiffness values. The average ΔV error for pickup trucks changed by 1 percent due to the average stiffness values remaining similar between versions of WinSMASH.

The 5- to 12-percent increases in stiffness for vans, utility vehicles, and cars resulted in substantial reductions in the ΔV error for these vehicles. Average ΔV error for vans and utility vehicles was reduced to 5.2 percent. The division of vans and utility vehicles from one stiffness category in WinSMASH 2.42 to four categories in WinSMASH 2008 provided more accurate representation for this diverse group of vehicles. The accuracy for car ΔVs was likewise improved and the underestimation was reduced to 16.0 percent on average. This was a great improvement but showed that there was still room to improve. The higher error for cars may possibly be due to cars having more stiffness variation due to larger crush or override by other vehicles due to bumper height mismatches.

The differences in ΔV error when using the vehicle specific stiffness values versus the updated categorical stiffness values was assessed by computing the ΔVs for a subset of vehicles using both methods. The two methods of obtaining vehicle stiffness values were found to be equal in terms of both ΔV accuracy and correlation to the EDR reported ΔV. This removed a potential source of variability and implied that the ΔV can be reasonably estimated for rare or extremely new vehicles with equal confidence to that of more common vehicles.

Aspects Consistent With Previous Versions

The extent of vehicle structural overlap was identified as having influence on the ability of WinSMASH to estimate ΔV, with the ΔV predictions being better for vehicles with major overlap of the damaged vehicle side. A study by Niehoff and Gabler (2006) showed that reconstructions of vehicles with extensive overlap were more accurate. In this study, overlap was assessed from the specific longitudinal location component of the CDC, which eliminated the other vehicle from the calculation and allowed for the overlap to be assessed for single vehicle crashes. The vehicles with direct damage to more than half of the damaged side underestimated the ΔV by 11.7 percent, which was 12.4 percent better than the vehicles with partial overlap. Because the vehicle stiffness values were obtained from crash tests that typically have full structural overlap, it was not surprising that these types of crashes can be reconstructed with more accuracy.

The recorded confidence for a reconstruction was found to have a strong effect on the correlation of WinSMASH ΔV predictions to the EDR ΔV predictions, but did not have an effect on the average error in ΔV. The correlation for the subgroup of low-confidence reconstructions was among the lowest of all subgroups examined in this study. Because only 23 percent of the crashes were low-confidence, the removal of these vehicles from the dataset resulted in only a moderate improvement in correlation and no change in the ΔV accuracy.

The accuracy of the ΔV for high-and low-ΔV crashes was examined through the use of arbitrary ΔV thresholds ranging from 16.1 – 32.2 kph (10 – 20 mph). The correlation between the WinSMASH and EDR ΔVs was much stronger for the high-speed group, regardless of what threshold was used. The correlations for the low-ΔV group were the lowest for any group and were observed to worsen as the ΔV threshold was lowered. The error in the low-speed group was attributed to the use of stiffness values derived from higher speed crash tests, assumptions that restitution was negligible, and the assumption that damage to the vehicle was purely plastic.

The errors in the average WinSMASH ΔVs were found to be dependent on the calculation type employed as well. The best correlations to the EDR-reported ΔVs were obtained when using the standard and barrier calculations, which were also the types that required the greatest amount of information about the vehicles. For crashes with limited information available, the missing vehicle calculation type offered a better correlation to the EDR ΔVs than the CDC only calculation type.

Implications for Current and Future Research

The changes to WinSMASH will result in a step change to the average ΔV between the NASS/CDS 2007 and earlier years and the newer years, making it more difficult to compare data across the two year ranges. However, as more data becomes available over the following years, the WinSMASH estimated ΔVs can be expected to become increasingly accurate.

The improvements in the ΔV estimates by vehicle body type have resulted in more consistent ΔV estimates across the body types, particularly for vans, pickup trucks, and utility vehicles. These changes will allow for more accurate comparisons of the merits and risks associated with each type of vehicle.

The increase in average ΔV in the new NASS/CDS case years due to the changes to WinSMASH may alter the interpretation of previous research such as studies of the relationship between ΔV and occupant injury, as well as the design and interpretation of crash tests where the impact speed was based on NASS/CDS ΔVs.

In future studies, it may be advisable to consider cases involving minimal overlap separately from cases with more substantial overlap. The results in Figure 9 and Figure 10 indicate that the level of engagement does make an observable difference in WinSMASH accuracy. A probable explanation for this is that WinSMASH frontal stiffness parameters are derived from full-engagement frontal crash tests, which are not at all representative of impacts engaging only the periphery of the frontal vehicle structure.

Limitations

The accuracy of the ΔVs reported in this study did not include the error inherent in the EDRs themselves. This error was reported to be 6 percent by Niehoff et al. in their 2005 study. Other potential sources of error were the crush measurements, principal direction of force, the assumptions inherent to the WinSMASH reconstruction software, and the simplified representation of vehicle stiffness.

The EDRs used in this study were all obtained from GM vehicles involved in moderate to severe frontal crashes. It is not known how these findings will generalize to other vehicle types or crash modes. The GM EDRs only record crashes with a longitudinal component sufficient to trigger air bag deployment. The dataset was not representative of all crashes in NASS/CDS.

Conclusions

A total of 469 crashes involving 478 vehicles experiencing no more than a single event and equipped with EDRs that recorded complete ΔV pulses were reconstructed using the enhanced WinSMASH reconstruction software. Because GM EDRs record only longitudinal ΔV, the dataset was composed of frontal and angled frontal impacts only.

Compared to the EDR maximum reported ΔV, the new version of WinSMASH underestimated the ΔV by 13 percent on average. This represented a large improvement over the 23 percent underestimation reported for the older WinSMASH 2.42.

The variability in ΔV estimates caused by the body type of the vehicles was greatly reduced in the enhanced WinSMASH. Pickup trucks and utility vehicles were all within 5 percent of the EDR reported ΔV on average. Vans underestimated the ΔV by 11 percent. All cars underestimated the ΔV by 16 percent, which was an improvement over the reported 27 percent underestimation for WinSMASH 2.42. CDC-only reconstructions consistently overestimated ΔV.

The accuracy of ΔV estimates was best for crashes with extensive overlap, high degrees of investigator confidence, and higher ΔVs, all of which was consistent with observations of previous versions of WinSMASH. ΔVs obtained using the new and updated categorical stiffness values were consistent with the ΔVs obtained using the new vehicle specific stiffness values. Because of the changes to WinSMASH, a step change in the average ΔV for NASS/CDS may be expected. However, the overall accuracy of the ΔV predictions will continue to improve as more crashes reconstructed with enhanced versions of WinSMASH (WinSMASH 2008 and onward) become available in databases, such as NASS/CDS.

References

Society of Automotive Engineers. (1980). Collision Deformation Code, SAE J224. Warrendale, PA: Society of Automotive Engineers International.

Eisenhauer, J. G. (2003). Regression Through the origin. *Teaching Statistics,* Vol. 25 (3), pp. 76-80.

Gabauer, D. J.,& Gabler, H. C. (2008). Comparison of roadside crash injury metrics Using event data recorders. *Accident Analysis & Prevention,* Vol. 40 (2), pp. 548-558.

Hampton, C. E.,& Gabler, H. C. (2009). NASS/CDS ΔV estimates: The influence of enhancements to the WinSMASH crash reconstruction code. *Annual Proceedings/Association for the Advancement of Automotive Medicine,* Vol. 53, pp. 91-102.

Lenard, J., Hurley, B., & Thomas, P. (1998, June). The accuracy of CRASH3 for calculating collision severity in modern European cars. *Proceedings of the 16th International Conference on Enhanced Safety of Vehicles,* Paper No. 98-S6-O-08, Windsor, Canada.

National Highway Traffic Safety Administration. (1986). CRASH3 Technical Manual (Report No. DOT HS 806 993). Washington, DC: National Highway Traffic Safety Administration.

Niehoff., P., Gabler., H. C., Brophy, J., Chidester, C., Hinch, J., & Ragland, C. (2005).Evaluation of event data recorders in full systems crash tests. *Proceedings of the 19th International Conference on Enhanced Safety of Vehicles*, Paper No. 05-0271-O.

Niehoff, P., & Gabler, H. C. (2006). The accuracy of WinSMASH ΔV estimates: The influence of vehicle type, stiffness, and impact mode. *Annual Proceedings/Association for the Advancement of Automotive Medicine*, Vol. 50, pp. 73-89.

Nolan, J. M., Preuss C. A., Jones, S. L., & O'Neill, B. (1998, June). An update on relationships between computed ΔVs and impact speeds for offset crashes. *Proceedings of the 16th International Conference on Enhanced Safety of Vehicles*, Paper No. 98-S6-O-07. Windsor, Canada.

O'Neill, B., Preuss, C. A., & Nolan, J. M. (1996, May). Relationships between computed ΔVs and impact speeds in offset crashes. *Proceedings of the 15th International Conference on Enhanced Safety of Vehicles,* Paper No. 98-S6-O-11. Melbourne, Australia.

Prasad, A. K. (1990). CRASH3 Damage Algorithm Reformulation for Front and Rear Collisions. (SAE Paper 900098) Warrendale, PA: Society of Automotive Engineers International.

Prasad, A. K.. (1991a). Missing Vehicle Algorithm (OLDMISS) Reformulation. (SAE Paper 910121). Warrendale, PA: Society of Automotive Engineers International.

Prasad, A. K. (1991b). Energy Absorbed by Vehicle Structures in Side Impacts. (SAE Paper 910599). Warrendale, PA: Society of Automotive Engineers International.

Sharma, D., Stern, S., Brophy, J., & Choi, E. (2007). An Overview of NHTSA's Crash Reconstuction Software WinSMASH. *The Proceedings of the 20th International Conference on the Enhanced Safety of Vehicles*. Paper Number 09-0211. Lyons, France.

Smith, R. A., & Noga, J. T. (1982). Accuracy and Sensitivity of CRASH. (SAE Paper 821169). Warrendale, PA: Society of Automotive Engineers International.

Stucki, S. L., & Fessahaie, O. (1998). Comparison of Measured Velocity Change in Frontal Crash Tests to NASS Computed Velocity Change. (SAE Paper 980649). Warrendale, PA: Society of Automotive Engineers International.

6. AutoSMASH

Introduction

This chapter describes AutoSMASH, an automated version of WinSMASH intended to compute ΔV for an entire case year in one run. AutoSMASH was originally developed as an internal research tool to support the WinSMASH accuracy study presented earlier in this report. Reconstructing the thousands of NASS/CDS cases used to examine the effects of WinSMASH 2010 would have been entirely impractical if the process were not automated. At the successful completion of that study, NHTSA requested that AutoSMASH be converted from a research tool to a production tool that could be used internally for automated quality control on NASS/CDS cases. The production version of AutoSMASH, described in this chapter, can be used both for internal research purposes and large-scale quality control by NHTSA.

Approach

AutoSMASH provides automated retrieval of reconstruction inputs directly from the NASS EDS(also referred to as the Oracle database), but can also use manually assembled input tables instead. Users also have a great deal of flexibility in how cases are reconstructed; AutoSMASH can reconstruct cases using any prior version of WinSMASH back to v2008.

Figure 17 depicts the operation of AutoSMASH. AutoSMASH gathers reconstruction data, e.g. crush measurements, from the NASS EDS, sends it to WinSMASH and then retrieves the results. NASSMAIN also uses WinSMASH to conduct reconstructions. However, while NASSMAIN deals with cases individually and depends on the user to operate WinSMASH, AutoSMASH processes many cases automatically in sequence and uses WinSMASH to obtain reconstructions without user interaction.

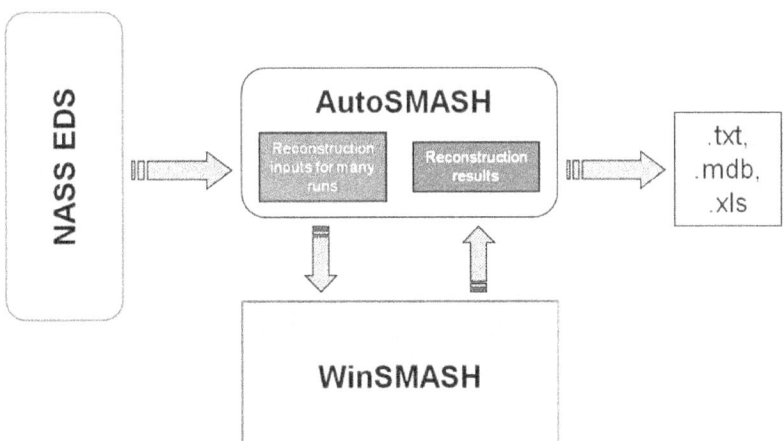

Figure 17. Diagram of the function of AutoSMASH.

AutoSMASH first assembles the data necessary to reconstruct a specified set of WinSMASH cases from a number of different tables within the NASS EDS.

AutoSMASH stores data internally as an Access database. Reconstruction inputs for each individual case are read from this internal database, and input to the appropriate version of WinSMASH. WinSMASH computes ΔV for the case and returns the results to AutoSMASH for storage in an output database. Once all the selected runs have been reconstructed, the results are assembled into an output table that may be saved in a number of formats.

No user interaction is required once AutoSMASH is started. The AutoSMASH-WinSMASH link uses a specialized code interface, distinct from the one normally used by NASSMAIN-WinSMASH. This interface allows AutoSMASH to command WinSMASH to reconstruct the sent case without displaying the WinSMASH form, or any error messages. This interface was developed for WinSMASH 2011.1.1.01, and retroactively ported to special builds of all prior WinSMASH versions to facilitate their use with AutoSMASH.

Operating AutoSMASH

Upon starting, AutoSMASH first displays the form shown in Figure 18. On this form, the user selects their source of reconstruction data. Users may choose either to reconstruct cases listed in an external table or may elect to reconstruct all cases corresponding to one or more case years directly from the NASS EDS. If the user selects input from an external table, the user must specify the location of a Microsoft Access database file (*.mdb) or Microsoft Excel spreadsheet (*.xls, *.xlsx) containing a table with the columns and data types enumerated in Appendix B (tables may also contain additional columns). If the user selects input from the NASS EDS, the user must provide their NASS EDS login credentials and select the case years that they wish to reconstruct. Note that only users with EDS access privileges may utilize this option.

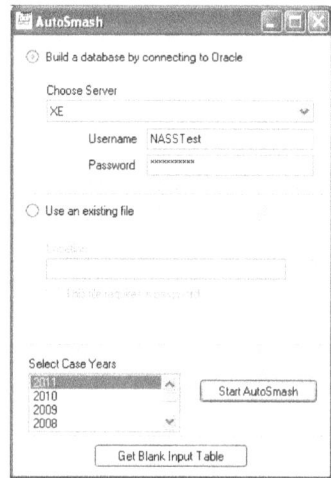

Figure 18. Initial screen of AutoSMASH.

Upon clicking "OK," the user is then presented with the main form of AutoSMASH (Figure 19). The main form allows users to select which version of WinSMASH should be used for the reconstruction. The main form also allows the user to inspect any input tables. All versions of WinSMASH detected by AutoSMASH appear in the Calculation Options area. The default option is to automatically reconstruct each case with the WinSMASH version corresponding to its case year. For example, under the default option, all NASS/CDS 2008 cases would be reconstructed using WinSMASH 2008.8.13.05, the highest version for 2008.

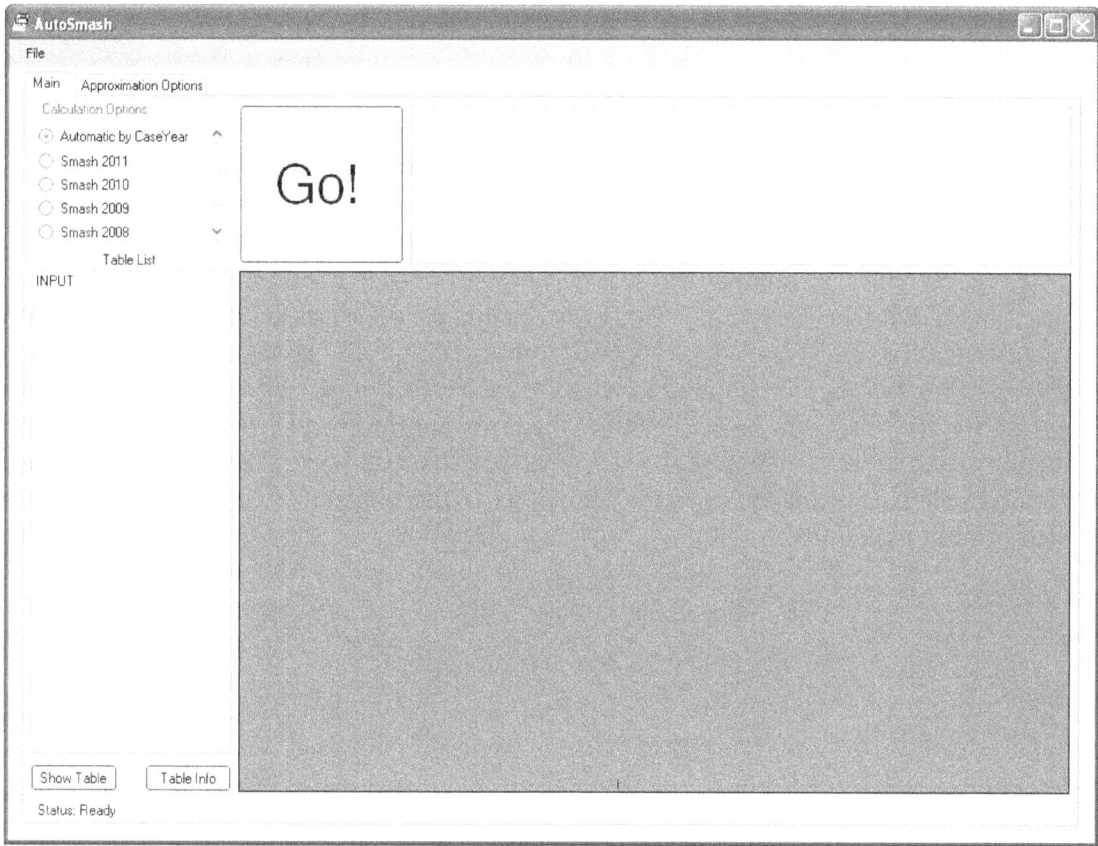

Figure 19. The main form of AutoSMASH.

There were a handful of bugfixes made between WinSMASH 2008 and 2009, as well as a stiffness database update. WinSMASH 2010 contains several bugfixes over WinSMASH 2009, as well as some differences in the way missing vehicles are handled and how damage is approximated for CDC-only vehicles. WinSMASH 2011 (v. 2011.1.1.01, the 2011 revision to WinSMASH 2010) also contains new bugfixes, a number of new features, and altered behavior when reconstructing missing vehicle crashes with side damage.

If the user elected to provide an input table, only this table will appear in the list; if the user elected to have AutoSMASH compile reconstruction inputs from the NASS EDS, then a number of intermediary tables will be listed in addition to the finished input table. Once the user is satisfied with the list of inputs, clicking "Go" will begin reconstruction of the assembled cases. Reconstruction may be stopped at any time by clicking this same button, which is labeled "Stop" while reconstructions are being run.

AutoSMASH Outputs

After reconstructing a set of cases using AutoSMASH, the results are stored in a table that is viewable via the Table List. This table contains all the same columns as the input table (see Appendix B), with the addition of columns containing the results of the reconstructions. Any vehicle specification fields that were approximated by WinSMASH will contain the approximated values. This output table may be saved as either a table in a Microsoft Access database file (*.mdb), a Microsoft Excel spreadsheet (*.xls) or as delimited text (*.txt) using a delimiter specified by the user.

In the event that WinSMASH should fail to calculate results for a case, all fields for that case in the output table will be blank, aside from those identifying it. The case will also be listed in the white box to the right of the Go button (illustrated in Figure 19); the user may right click on a case number and select "Show Errors" to obtain more detailed information on why the case failed. This re-loads that particular case into WinSMASH and calls that version of WinSMASH's error-checking code. WinSMASH will then display all error messages pertaining to why the case failed to reconstruct. Because it is WinSMASH itself that runs error checking, the error-checking and error messages will be a function of the version of WinSMASH with which the case is checked. Newer versions of WinSMASH have been equipped with much more thorough error checking than was present in earlier versions.

7. Summary

This project has developed WinSMASH 2010, a completely rewritten and restructured version of the WinSMASH crash reconstruction code used in conjunction with the NASS EDS (aka Oracle). The new version of the code has corrected known programming bugs, implemented an improved strategy for the lookup of stiffness values, rewritten the code in the more modern language C#, and developed an expanded library of stiffness values that better reflect the current vehicle fleet. Earlier phases of this project developed WinSMASH 2007 and WinSMASH 2008. WinSMASH 2007 was used to generate ΔV estimates in NASS/CDS 2007. WinSMASH 2008 was used for the same function in NASS/CDS 2008. WinSMASH 2010 generated ΔV estimates in NASS/CDS 2009 and NASS/CDS 2010 and will be used to generate ΔV estimates for NASS/CDS 2011 and later.

The influence of these changes was analyzed by recomputing the entire year of NASS/CDS 2007 cases using WinSMASH 2008, and comparing the results with existing ΔV values in NASS/CDS 2007 computed using WinSMASH 2007. This analysis has shown that the use of WinSMASH 2008 will increase the average ΔV for a NASS/CDS year by 8.6 percent. This corresponded to a 1.8 kph (or 1.1 mph) increase on average for each vehicle. The error with respect to EDR measured ΔV has dropped from 23 percent underestimation to 13 percent underestimation on average.

The use of WinSMASH 2010 will have two important implications. First, the analysis suggests that the use of WinSMASH 2010 will help to alleviate the underestimation of ΔV reported in previous research. Second, the change in average ΔV could result in a discontinuity in ΔV between NASS/CDS 2008 and the earlier years. This discontinuity varies by the general area of damage. Specialized analytical techniques may be required to allow aggregation of NASS/CDS 2008 and later with earlier years of NASS/CDS.

8. References

National Highway Traffic Safety Administration. (1986). CRASH3 Technical Manual (Report No. DOT HS 806 993). Washington, DC: National Highway Traffic Safety Administration.

Niehoff, P., & Gabler, H. C. (2006). The accuracy of WinSMASH ΔV estimates: The influence of vehicle type, stiffness, and impact mode. *Annual Proceedings/Association for the Advancement of Automotive Medicine*, Vol. 50.

Prasad, A. K. (1990). CRASH3 Damage Algorithm Reformulation for Front and Rear Collisions. (SAE Paper 900098) Warrendale, PA: Society of Automotive Engineers International.

Prasad, A. K.. (1991a). Missing Vehicle Algorithm (OLDMISS) Reformulation. (SAE Paper 910121). Warrendale, PA: Society of Automotive Engineers International.

Prasad, A. K. (1991b). Energy Absorbed by Vehicle Structures in Side Impacts. (SAE Paper 910599). Warrendale, PA: Society of Automotive Engineers International.

Sharma, D., Stern, S., Brophy, J., & Choi, E. (2007). An Overview of NHTSA's Crash Reconstuction Software WinSMASH. *The Proceedings of the 20th International Conference on the Enhanced Safety of Vehicles*. Paper Number 09-0211. Lyons, France.

Appendix A – Revisions to WinSMASH

Corrections of Programming Bugs in WinSMASH v2.44

- The Pole option is disabled.
- The values for vehicle damage length (SmashL) are no longer reset to zero after a case is run or the case is saved to a CR4 file.
- For vehicle 2, on the crush tab, the value for C1 no longer disappears, and the value for C3 remains in the proper entry box. If vehicle 2 has more than two crush points entered, the calculation will use all of the available values as opposed to only the first two.
- A bug was identified in version 2.44 that caused WinSMASH to invariably use an approximated front overhang value, even when the user explicitly supplied a value. This has been rectified.
- Previously, WinSMASH would allow calculations to proceed without a damage length specified (for non-CDC-only vehicles). This has been rectified.
- WinSMASH now properly reads cases from CR4 files where Vehicle 1 is indicated to have rotated more than 360 degrees.
- WinSMASH no longer uses crush values from previous runs in new missing vehicle calculations.
- Trajectory simulation results for vehicle 2 can now be viewed, whereas before they could not.
- WinSMASH now properly displays the weight distribution ("Percent Front") for all vehicles.
- WinSMASH will no longer allow users to view the results of saved cases that load incorrectly but have calculated results. In order to view the results tab, the case must have the correct run type and all necessary information entered. The case is recalculated regardless of status when the results are viewed. This primarily affects files saved with older versions of WinSMASH.
- For missing vehicle cases, the PDOF of the missing vehicle is now automatically updated based on the data that the user enters for the two vehicle headings and the known vehicle PDOF. The missing vehicle's PDOF field is also disabled and will not allow the user to manually change the value.
- The values for damage length, Direct +/- D and Field L +/- D are now checked for damage and PDOF specified outside of the bounds of the vehicle. If a wheelbase or size category is defined and the user entered PDOF or damage length is not valid, then the user is not allowed to leave the entry box. If the values in question are loaded from a file, a warning is shown. If damage length or PDOF is specified before wheelbase or size category, the user is warned that damage length cannot be validated.
- WinSMASH now checks the PDOF against the specified damage side for each vehicle as it is entered.
- WinSMASH now checks vehicle colinearity when a case is run. If the colinearity check fails, WinSMASH suggests a new value for the heading or PDOF for vehicle 2.
- WinSMASH no longer generates "Invalid Size Category" errors if the user attempts to leave the Wheelbase, Front Overhang, or Weight Distribution fields without entering a non-zero value.

- In version 2.44, longitudinal values for "Speed Change: Momentum and Spinout" were assigned to the latitudinal (now lateral) field, and vice versa – this has been rectified.
- The trajectory calculations in version 2.44 never worked properly. Thus, they were replaced with a direct translation of the original FORTRAN CRASH3PC trajectory calculations into C#. Aside from several small bugfixes, the algorithm itself remained essentially unchanged.
- The CRASH3PC version of the trajectory calculations used a value of 384.6 (in/s^2) for gravitational acceleration. The correct value is 386.4 (in/s^2), and this is used in the C# translation of the CRASH3PC trajectory calculations. This had an effect on the values used for several vehicle parameters, most notably mass and radius of gyration.
- For certain collisions, the trajectory algorithm applies a correction to the vehicle headings at separation (see CRASH3 Technical Manual, section 2.3.4, page 2.68). In CRASH3PC, this correction would only be applied if the vehicles rotated clockwise when it should be applied regardless of the direction of rotation. This has been rectified in the C# translation.

New Features

Main form
- For barrier, missing or CDC-only cases, either vehicle 1 or vehicle 2 may now be the missing/barrier/pole/CDC vehicle.
- A "Print All" menu option has been implemented.
- The user may now select which printer they wish to print to.
- WinSMASH can now "print" runs to a .pdf file without third-party software present on the host machine.
- All printed pages now have a header and footer that indicate the version of WinSMASH used to print the run, the version of the database file associated with WinSMASH, the date and time the printout was printed, the name of the .cr4 file from which the run came (if applicable) and the page number.
- The Review tab now shows the columns "Final" and "Valid" for the standalone version of WinSMASH as well as the DLL.
- Users may now obtain a list of all differences between two runs on the Review grid. One run is selected, a new "Compare To:" button is clicked, and a second run is selected. Any differences between the runs are displayed in dialog boxes.
- Values on the Specifications page are now color-coded as to their origins: black text indicates a user-entered value, green indicates an approximation (including categorical stiffnesses) and blue indicates a vehicle-specific value retrieved from the WinSmash vehicle database.
- There are now labels under "Damage Length" and "Damage Offset" to indicate the name of these fields in the NASS system (SMASH L and Direct +/- D respectively).
- Checkboxes have been implemented for every vehicle specification that can be approximated. These checkboxes allow the user to explicitly control which fields WinSMASH approximates and which it does not, eliminating much of the complicated protocol that previously governed this.
- Vehicle heading input values are now restricted to multiples of 5 degrees.

- The "Get Vehicle Specifications" button has been changed to the "Get Vehicle Stiffness" button. In normal operation, instead of opening the search window WinSMASH will now automatically retrieve the matching vehicles from the database. There are several possible outcomes:
    - One match with stiffnesses defined: WinSMASH will retrieve the stiffnesses and put them into the d0 and d1 fields.
    - No matches: WinSMASH will select a stiffness category based on wheelbase, bodystyle and damaged side.
    - Multiple matches: WinSMASH will ask the user to check that they have entered a year, make, model, and bodystyle.
- Vehicle stiffness can still be manually edited/selected by explicitly enabling a new "Advanced" option – otherwise, the stiffness is handled via the new "Get Vehicle Stiffness" routine.
- With "Advanced" mode activated, values entered in the vehicle Stiffness Category will override all other methods of entering stiffness data, so long as the Stiffness Category is not zero. This means that user-entered values for d0 and d1 will be lost if the Stiffness Category value for that vehicle is non-zero.
- WinSMASH now restricts the user from running calculations without first retrieving the latest stiffness values from the database (unless Advanced mode is activated).
- The "Get Vehicle Stiffness" button does not by default retrieve vehicle specs other than stiffness – this can still be done optionally with Advanced mode activated.
- The vehicle database may still be searched in Advanced mode, and database queries may now be made with only partial vehicle information.
- The Stiffness Category is now entered by choosing from a drop down list (when it is manually entered in Advanced Stiffness mode).
- A full help file has been integrated into the program. Users may browse it directly, or press F1 while hovering over a control on the form for help on a specific feature.
- The Help menu now contains a "Why Can't I Run Calculations?" feature. When selected, will inform the user exactly why the Calculate button is disabled (if it in fact is).
- It is now possible to toggle the units used in the display between English and Metric at the press of a button.

Trajectory Simulations
- The entire trajectory simulation interface has been remodeled; all forms related to trajectory simulation have been consolidated into a single window with four tabs. Each vehicle has its own tab that displays the results for that vehicle, the trajectory options window has been given its own tab and the fourth tab displays regular results that have been recalculated using the results of the simulation.
- In version 2.44, clicking the "Vehicle 1"/"Vehicle 2" button on the trajectory form would perform a new run for a given vehicle – this meant that looking back and forth between the simulation results for each vehicle would add new simulation iterations. This is no longer the case – the button that runs simulations is now used solely for that purpose, and viewing results for each vehicle is accomplished by clicking on the respective vehicle tabs.
- Users can either manually run how many ever simulation iterations they wish by clicking "Run Simulation" in the Trajectory form tab for a given vehicle, or WinSMASH can

automatically run iterations for both vehicles until the results have converged (error <= 0.001). The latter is accomplished by running an animation from the Graphics tab: if no simulations are already present, WinSMASH will run simulations until the results converge satisfactorily (again, error <= 0.001), and these results will then be shown in the Trajectory window. Non-convergent simulations are capped at 300 iterations.
- Trajectory simulation data is now cleared when a new run is loaded or when some other information is altered in such a way that existing simulations might no longer be valid (this was not the case in version 2.44).
- The Trajectory form now has its own print button.
- The Help menu now contains a "Why Can't I Run Trajectory Analysis?" feature. When selected, will inform the user exactly why the option to simulate vehicle trajectory is disabled (if it in fact is).

Graphics
- To prevent screen clutter, a number of buttons were added to the bottom of the graphics window that allow the user to turn on or off all of the new display options. These include:
  - Vehicle-local X and Y axes;
  - The centroid of the vehicle damage area;
  - The vehicle wheel axles; and
  - The vehicle rotation direction. This rotation is based on the direction specified in the motion tab - the results for Omega (based on damage-only calculations) have no influence on this.
- WinSMASH now calculates the lengthwise portion of the damage centroid location in addition to the depth. This is used for drawing the damage centroid on the graphics tab, which was not possible in version 2.44.
- The PDOF indicator has been changed to an arrow; the text "PDOF" no longer appears on the graphics form.
- The PDOF arrow is now drawn on the center of *direct* damage. It will usually not coincide with the centroid locator, which indicates the centroid of *total* damage.
- Each vehicle has "V1" or "V2" drawn on it to indicate which vehicle is which. The full labels also specify "Vehicle 1" or "Vehicle 2" and the year, make, model and bodystyle.
- A Print button has been added to the graphics form. This button allows any of the Graphics tabs to be printed to the selected printer.
- Modifications to vehicle damage, heading, PDOF, etc, on the Main form will cause the Graphics form to refresh, allowing the user to easily see the effects of changes to case parameters.
- The Graphics Window can now be viewed at any time. This assists the user with vehicle positioning, as the vehicle positions on the screen update as the data is entered.
- Animation speed can now be adjusted to the user's preference.
- Vehicle graphics now render with the X-axis pointing up and the Y-axis pointing right, making them consistent with graphics in NASSMAIN.

Refinements

- WinSMASH has been rewritten in C#. C# offers the benefit of a widespread and well-documented development environment. Previously, WinSMASH was written in Delphi,

which is no longer supported by its parent company Borland. Being a relatively new language, C# is also in a good position to be optimized for the latest in computer hardware and operating systems.
- WinSMASH is now compiled in such a way that it is fully compatible with the latest operating systems, including 64-bit editions of Microsoft Windows 7.
- All closed-source third party software components in WinSMASH have been removed from or replaced with native components. This modification simplifies the licensing issues associated with WinSMASH distribution, and will make future code maintenance significantly more straightforward.
- The WinSMASH DLL now incorporates a code interface that allows "drop-in" use with the "AutoSMASH" quality control tool. New versions of WinSMASH may thus be used with AutoSMASH by simply providing a copy of the new WinSMASH DLL (and its required files).
- The code involved in closing the graphics window has been altered to improve application stability.
- WinSMASH will now check at startup for the vehicle database file. If it is not found, then WinSMASH will terminate with a message indicating why.
- Save prompts have been streamlined. WinSMASH now only prompts to save when changes have actually been made to a run.
- The vehicle database has been updated with new vehicle information, including new vehicle-specific stiffnesses.
- Stiffness category definitions and size category definitions are now stored entirely within the AVehicle database. WinSMASH 2.44 had category definitions directly defined in the code in multiple places, which led to different category definitions being used in different situations. By having all category definitions stored in one place external to the code, the definitions are consistently applied, and may be updated without requiring a rebuild of WinSMASH itself.
- The vehicle stiffness database was updated from a Paradox database to a Microsoft Access 2003 database, and the code for accessing the database was updated to reflect this.
- The code comprising the printing infrastructure has been centralized and streamlined to facilitate future enhancements and maintenance.
- Unit conversion factors have been consolidated into a single static class, thus providing a single, central point where the values are defined. Previously, conversions were hard-coded wherever they were used; WinSMASH had slightly different conversion factors for the same conversion in several places before this modification.
- The flickering of tabs when WinSMASH is launched has been eliminated.
- Selection of calculation type is now done by WinSMASH based on the information received from NASSMAIN.
- The "Size Category" field is no longer visible, as the size category is now selected automatically based on the wheelbase.
- The stiffness categories have been updated to reflect a more current vehicle fleet.
- In Missing vehicle cases with side damage to one of the vehicles, categorical stiffnesses are the default choice, while vehicle-specific stiffnesses will be given priority in all other instances. Missing cases with side damage were found to give better results with categorical stiffnesses.

- Previously, WinSMASH would not automatically fill in the d0 value for missing vehicles. This value is not needed for missing vehicles, but it helps to complete the form. Code has been added to ensure that WinSMASH fills in this value for all missing vehicles.
- Blank or negative crush values are now allowed. WinSMASH interprets these as meaning that no crush measurement was taken.
- The WinSMASH "Scene" tab and the "End Rotation" fields on the "Motion" tab now accept decimal values for all fields on the tab. The fields are also initialized to 0.00 to imply to the user that decimal inputs are acceptable.
- The radius of gyration now automatically updates when vehicle length is changed – this was previously not the case.
- On the Results tab, all labels saying "Latitudinal" now read "Lateral".
- The ΔV values shown under the columns "Speed Change: Momentum and Spinout" and "Impact Speed: Momentum and Spinout" have been swapped. The "Momentum" PDOF field remains in its original position, however.
- The Review tab now expands to fill the size of the form.
- When using the Review tab to load old cases, the PDOF now loads before the new stiffnesses. This prevents partial loading of cases and erroneous error messages regarding the validity of the stiffness values.
- WinSMASH now calculates the weight distribution (Percent Front field) differently than in previous versions. To avoid destroying results generated by previous versions of the program, user-entered weight distributions are not altered by WinSMASH. WinSMASH will however replace weight distribution values that are known to be approximated.
- The Direct +/- D and Field L +/- D measurements are referenced to the vehicle CG before use in calculations, rather than simply being used as-entered on the form (in which case they reference the center of wheelbase). The results for vehicles with R or L damage sides will thus be different from previous WinSMASH versions. This affects Omega and Barrier Equivalent Speed for the most part. Changes to the calculated Delta V are slight and usually imperceptible as displayed on the form. However, all drawings/animations are still made with measurements from the center of wheelbase.
- Code has been added to allow for different vehicle images to be displayed on the graphics tab for vehicles of different body types. As soon as images for new body types are created, the infrastructure is in place for them to be rendered.
- The positions of the vehicles on the graphics form now reflect the true crash configuration to allow for visual validation of the setup.
- Field L +/- D now positions the damage profile on the vehicle on the Graphics form instead of the Direct +/- D field.
- The "About" window is now coded such that it automatically detects the program and database version, meaning new application versions will always have correct version information displayed (assuming the developer increments the build number in the IDE for new versions).

Distribution to NHTSA Crash Investigators

- WinSMASH is now distributed via InstallShield installation packages to streamline the installation process and ensure correct integration with the NHTSA EDS software components that WinSMASH requires to function.
- The installer checks to see if the Microsoft .NET Framework is present, as this is necessary for WinSMASH to function. InstallShield has the capability to bundle the .NET redistributable with the installer, but this increases the size of the installer file greatly. Thus, if the .NET framework is not found, the user is prompted to install it themselves at present.
- For the DLL version of WinSMASH, the installer handles the COM registration and unregistration of the DLL automatically. The DLL version of WinSMASH makes use of "registration-free COM," which allows the DLL to be installed correctly without permission to edit the registry. This makes the program installation much easier to perform on heavily secured systems where the user is unlikely to have administrative permissions.
- Installation now creates a desktop shortcut and associates .CR4 files with WinSMASH, allowing .CR4s to be opened by double clicking on them.

Appendix B – AutoSMASH Input/Output Table Format

Table 13 lists the columns and data types that must be present in a user-supplied AutoSMASH input table, and additional columns that are contained in saved output from AutoSMASH. When supplying input, it is not necessary to provide any values for the result fields, e.g. coded_TotalDeltaV(1/2), coded_LongDeltaV(1/2), coded_LatDeltaV(1/2), coded_EnergyDeltaV(1/2) and coded_BarrierDeltaV(1/2). Additional fields may be left blank as well, depending upon the needs of WinSMASH when reconstructing a particular case. Table 14 gives the numerical codes to be used for specifying different calculation types in WinSMASH. Note that the specified data type requirements only apply to Microsoft Access input files; Microsoft excel spreadsheets must contain all the same column names, but the column format is irrelevant so long as any entered data can be parsed as the required type by AutoSMASH. Input tables may contain other columns in addition to the ones specified in Table 13, and required columns may be placed in any order.

Table 13. Column names, data types and units of all fields required for manually supplying cases to AutoSMASH via Microsoft Access or Microsoft Excel. Columns that indicate "output" in the description appear only in saved results, and are not required in input tables.

Field Name	Access Data Type	Units	Description
CaseYear	Long Integer	[years]	NASS case year
PSU	Long Integer	-	Primary Sampling Unit of case
CaseNumber	Long Integer	-	NASS case number
CaseID	Long Integer	-	CaseYear-PSU-CaseNumber
EventNo	Long Integer	-	NASS event number
CalcType	Long Integer	-	WinSMASH calculation type - see Table 14
SmashVersion	Text	-	WinSMASH version used in reconstruction
ModelYear1	Long Integer	[years]	v. 1 model year
MakeName1	Text	-	v. 1 make name
MakeID1	Long Integer	-	v. 1 NASS make ID code
ModelName1	Text	-	v. 1 model name
ModelID1	Long Integer	-	v. 1 NASS model ID code
Bodystyle1	Text	-	v. 1 bodystyle
CDC1	Text	-	v. 1 CDC
DamageSide1	Text	-	v. 1 damaged side ('F,' "B," "L," "R" or blank)
DirectionOfForce1	Long Integer	[deg]	v. 1 PDOF
HeadingAngle1	Long Integer	[deg]	v. 1 heading angle
Wheelbase1	Long Integer	[cm]	v. 1 whee base
OverallLength1	Long Integer	[cm]	v. 1 length
MaxWidth1	Long Integer	[cm]	v. 1 width
TotalWeight1	Long Integer	[kg]	v. 1 weight
FrontOverhang1	Long Integer	[cm]	v. 1 front overhang
CGLocation1	Long Integer	[cm]	v. 1 CG location (aft of front of vehicle)
Radius1	Long Integer	[cm]	v. 1 radius of gyration
PercentFront1	Double	[%]	v. 1 weight distribution
Size1	Long Integer	-	v. 1 WinSMASH size category - used only by early versions
Stiffness1	Long Integer	-	v. 1 WinSMASH stiffness category - optional

Field	Type	Units	Description
d0_1	Double	[sqrt(N)]	v. 1 d0
d1_1	Double	[sqrt(N/cm)]	v. 1 d1
SmashL1	Long Integer	[cm]	v. 1 Smash L (direct + induced damage length)
SmashD1	Long Integer	[cm]	v. 1 Direct +/- D (damage offset)
D1	Long Integer	[cm]	v. 1 Field L +/- D
C1_1	Long Integer	[cm]	v. 1 crush point
C2_1	Long Integer	[cm]	v. 1 crush point
C3_1	Long Integer	[cm]	v. 1 crush point
C4_1	Long Integer	[cm]	v. 1 crush point
C5_1	Long Integer	[cm]	v. 1 crush point
C6_1	Long Integer	[cm]	v. 1 crush point
ModelYear2	Long Integer	[years]	v. 2 model year
MakeName2	Text	-	v. 2 make name
MakeID2	Long Integer	-	v. 2 NASS make ID code
ModelName2	Text	-	v. 2 model name
ModelID2	Long Integer	-	v. 2 NASS model ID code
Bodystyle2	Text	-	v. 2 bodystyle
CDC2	Text	-	v. 2 CDC
DamageSide2	Text	-	v. 2 damaged side ('F," "B," "L," "R" or blank)
DirectionOfForce2	Long Integer	[deg]	v. 2 PDOF
HeadingAngle2	Long Integer	[deg]	v. 2 heading angle
Wheelbase2	Long Integer	[cm]	v. 2 whee base
OverallLength2	Long Integer	[cm]	v. 2 length
MaxWidth2	Long Integer	[cm]	v. 2 width
TotalWeight2	Long Integer	[kg]	v. 2 weight
FrontOverhang2	Long Integer	[cm]	v. 2 front overhang
CGLocation2	Long Integer	[cm]	v. 2 CG location (aft of front of vehicle)
Radius2	Long Integer	[cm]	v. 2 radius of gyration
PercentFront2	Double	[%]	v. 2 weight distribution
Size2	Long Integer	-	v. 2 WinSMASH size category - used only by early versions
Stiffness2	Long Integer	-	v. 2 WinSMASH stiffness category - optional
d0_2	Double	[sqrt(N)]	v. 2 d0
d1_2	Double	[sqrt(N/cm)]	v. 2 d1
SmashL2	Long Integer	[cm]	v. 2 Smash L (direct + induced damage length)
SmashD2	Long Integer	[cm]	v. 2 Direct +/- D (damage offset)
D2	Long Integer	[cm]	v. 2 Field L +/- D
C1_2	Long Integer	[cm]	v. 2 crush point
C2_2	Long Integer	[cm]	v. 2 crush point
C3_2	Long Integer	[cm]	v. 2 crush point
C4_2	Long Integer	[cm]	v. 2 crush point
C5_2	Long Integer	[cm]	v. 2 crush point
C6_2	Long Integer	[cm]	v. 2 crush point
dvRank1	Long Integer	-	NASS "Ranking" for existing v. 1 ΔV
coded_TotalDeltaV1	Long Integer	[kph]	v. 1 existing total ΔV
new_TotalDeltaV1	Long Integer	[kph]	output - v. 1 total ΔV
coded_LongDeltaV1	Double	[kph]	v. 1 existing longitudinal ΔV
new_LongDeltaV1	Double	[kph]	output - v. 1 longitudinal ΔV

coded_LatDeltaV1	Double	[kph]	v. 1 existing lateral ΔV
new_LatDeltaV1	Double	[kph]	output - v. 1 lateral ΔV
coded_EnergyDeltaV1	Long Integer	[J]	v. 1 existing absorbed energy
new_EnergyDeltaV1	Long Integer	[J]	output - v. 1 absorbed energy
coded_BarrierDeltaV1	Double	[kph]	v. 1 existing BES
new_BarrierDeltaV1	Double	[kph]	output - v. 1 BES
dvRank2	Long Integer	-	NASS "Ranking" for existing v. 2 ΔV
coded_TotalDeltaV2	Long Integer	[kph]	v. 2 existing total ΔV
new_TotalDeltaV2	Long Integer	[kph]	output - v. 2 total ΔV
coded_LongDeltaV2	Double	[kph]	v. 2 existing longitudinal ΔV
new_LongDeltaV2	Double	[kph]	output - v. 2 longitudinal ΔV
coded_LatDeltaV2	Double	[kph]	v. 2 existing lateral ΔV
new_LatDeltaV2	Double	[kph]	output - v. 2 lateral ΔV
coded_EnergyDeltaV2	Long Integer	[J]	v. 2 existing absorbed energy
new_EnergyDeltaV2	Long Integer	[J]	output - v. 2 absorbed energy
coded_BarrierDeltaV2	Double	[kph]	v. 2 existing BES
new_BarrierDeltaV2	Double	[kph]	output - v. 2 BES
ID	Long Integer	-	primary key, no duplicates

Table 14. Numerical codes for WinSMASH calculation types.

WinSMASH Calculation Type	Numerical Code
Standard	1
Missing – v1	6
Missing – v2	3
CDC Only – v1	7
CDC Only – v2	5
Barrier – v1	8
Barrier – v2	2

Appendix C – AutoSMASH Programmer's Guide

Support for Previous WinSMASH Versions

One of the novel features of AutoSMASH is the ability to run reconstructions using versions of WinSMASH as early as WinSMASH 2008. The default option is to automatically reconstruct each case with the WinSMASH version corresponding to its case year; the user may opt to reconstruct all runs from each selected case year using one version of WinSMASH. The following table presents the version of WinSMASH that will be chosen by default for case reconstruction. Note that WinSMASH versions prior to 2008 are not available for use with AutoSMASH, and hence, earlier years of NASS/CDS will default to the earliest available version of WinSMASH, WinSMASH 2008.

Table 15. AutoSMASH Default versions of WinSMASH

NASS/CDS case year	Default WinSMASH version
pre-2007	WinSMASH 2008.8.13.05
2007	WinSMASH 2008.8.13.05
2008	WinSMASH 2008.8.13.05
2009	WinSMASH 2008.11.24.04
2010	WinSMASH 2010.6.2.00
2011	WinSMASH 2011.1.1.01

AutoSMASH includes specialized copies of these earlier versions of WinSMASH, as well as their accompanying WinSMASH stiffness libraries. These copies are specialized because they have all been recompiled to include a special code interface used by AutoSMASH, which allows calculations to be run without displaying the form or showing messages. This interface was not present in WinSMASH prior to version 2011.1.1.01. All new WinSMASH versions will be compatible with AutoSMASH, but only the specialized copies of prior versions are compatible.

AutoSMASH itself searches for folders within its own root directory named for a year (i.e., folders "2007," "2008," "2010," etc.) and ignores folders whose names are not years (i.e., folders "2008.11.24.04," "2010 debug," "temp," etc.). AutoSMASH then scans each appropriately named folder for WinSMASH DLLs implementing the software interface described above. AutoSMASH will then associate the WinSMASH DLL within each directory with that particular case year.

If multiple eligible DLLs are found within the same folder, the copy with the highest (largest) version number will be used. New versions of WinSMASH can thus be added to AutoSMASH by simply creating a new directory and inserting copies of the DLL and vehicle database: AutoSMASH will detect them and add them to the list of WinSMASH version choices within the program automatically. Again, all that is required is that the folder name be a year and that the DLL implement the software interface.

Automated Retrieval of Case Information from Oracle

Collection of the data from the Oracle (NASS EDS) data tables necessary to reconstruct cases is a challenging procedure, requiring data to be extracted from more than seven different tables. Because cases can have multiple events, and involve multiple vehicles, great care is taken to ensure that each set of reconstruction inputs is matched to the correct vehicle, and that in each event the correct set of vehicles is involved. Not every piece of information necessary to fully recreate a WinSMASH reconstruction is recorded in the Oracle EDS database. Some WinSMASH inputs (CalcType, Size, Stiffness) must be inferred from what is explicitly stored. Information from the WinSMASH tables was not used due to the concerns about the Valid and Final tags and the possibility of case reviews resulting in changes to the Oracle data that would not be reflected in the WinSMASH tables (discussed at the end of this section)

Tables Used

The major Oracle tables containing the data needed by WinSMASH are shown below in Table 16. The tables VEHICLE, EVENT, and OCCUPANT contain general information about the undamaged vehicle specifications, event, and vehicle occupants respectively. The DAMAGELOCATOR table contains all crush profiles for those vehicles for which measurements were made. The vehicle results are split into two different tables. VEHICLECRASH contains information for vehicles that were not inspected: these vehicles were unavailable for inspection and typically do not have an associated crush profile. VEHICLEDEFORMATION contains the results for inspected vehicles, as well as information used to generate a CDC code for the vehicle. Since cases are stored in NASS by the CaseID field, the SELECTED table is necessary to identify the year, PSU and case number for each case.

Table 16. The Oracle tables queried by AutoSMASH, and the relevant fields.

Table	Fields
DAMAGELOCATOR	CaseID, VehicleID, ImpactID, ImpactNumber, CDCID, SmashL, SmashD, D, Avg_C1, Avg_C2, Avg_C3, Avg_C4, Avg_C5, Avg_C6
VEHICLECRASH	CaseID, VehicleID, CollisionID, DeltaVEventID, HeadingAngle, HighestDeltaV, HighestLongDeltaV, HighestLatDeltaV, HighestEnergy, HighestBarrierSpeed
VEHICLEDEFORMATION	CaseID, VehicleID, DeformID, EventSequenceID, Ranking, DirectionOfForce, HeadingAngle, ClockForce, DeformationLocation, LongLatLocation, VertLatLocation, DamageDistribution, DamageExtent, PickDeltaVBasis, TotalDeltaV, LongDeltaV, LatDeltaV, EnergyDeltaV, BarrierDeltaV, ConfidenceID
VEHICLE	CaseID, VehicleID, VehicleNumber, ModelYear, MakeID, ModelID, BodytypeID, Wheelbase, OverallLength, MaxWidth, CurbWeight, CargoWeight, FrontOverhang
SELECTEDCASES	CaseID, CaseYear, PSU, CaseNumber
EVENT	CaseID, VehicleID, EventSequenceID, PickAreaOfDamage, PickContactVehicleDamage
OCCUPANT	CaseID, VehicleID, OccupantID, OccupantNumber, Weight, Age, Sex

The tables listed in Table 16 supply the raw information required for WinSMASH reconstructions. However, additional tables are required to map many of the numerical codes in these to their corresponding strings, or to obtain standard approximations for unlisted values. Tables used for such purposes:

- CDCLOOKUP
- BODYTYPE_X_WINSMASH
- MAKELOOKUP

- MODELLOOKUP
- OCCWEIGHTLOOKUP

Combining Vehicle and Occupant Data

The first step in retrieving inputs for a case is to combine the general information about the vehicle and its occupants into a form useable by WinSMASH. In particular, the many weight fields (vehicle, cargo, occupants) must be condensed into a single total weight value for the vehicle. One complication of this is that the weight for each of the occupants is not always stored in the Oracle database. In such situations, a representative occupant weight is assigned from the OCCWEIGHTLOOKUP table based on the age and sex of the occupant. Another noteworthy step is the assignment of make, model and bodystyle names from MAKELOOKUP, MODELLOOKUP and BODYTYPE_X_WINSMASH. All versions of WinSMASH 2008 earlier require the make and model names of a vehicle to perform stiffness parameter lookups, so the vehicle make and model ID numbers are used to assign the proper text to the make and model name fields. WinSMASH 2010 and later will use the IDs directly, even in the event that strings are passed, and WinSMASH 2007 uses only categorical vehicle stiffnesses (and so performs no vehicle lookups). Once assembled, this vehicle and occupant information is merged together.

Combining Results for Inspected and Uninspected Vehicles

The next step is to collect the reconstruction results, which are split between the VEHICLECRASH and VEHICLEDEFORMATION tables. The table VEHICLEDEFORMATION contains all of the results for inspected vehicles while the table VEHICLECRASH contains results for uninspected vehicles. A field called PickDeltaVBasis indicates whether WinSMASH was used to provide ΔV estimates; any case not reconstructed with WinSMASH is discarded, leaving only pertinent records.

Extraction of records from VEHICLECRASH is somewhat more complicated. VEHICLECRASH contains a large number of records for uninspected vehicles; however, many of these records are blank or recorded as unknown ΔV. These blank/unknown records are eliminated by selecting only records for which the HighestDeltaV and DeltaVEventID fields are greater than zero.

The VEHICLECRASH table contained up to two sets of results. The fields for the first set represented the highest severity reconstructed event and were prefixed with "HIGHEST." Similarly, the second set contained the second highest severity reconstructed event and prefixed with "SECOND." Each row in VEHICLECRASH was split into two rows with one set of results. The names of these results fields were changed to match those in VEHICLEDEFORMATION as shown in Table 17. We added a field named "RANKING" was for the purpose of tracing the results back the "HIGHEST" or "SECOND" field groups if needed.

Table 17. VEHICLECRASH field name changes.

VEHICLECRASH name	VEHICLEDEFORMATION name
DeltaVEventID, SecondDeltaVEventID	EventSequenceID
HighestDeltaV, SecondDeltaV	TotalDeltaV
HighestLongDeltaV, SecondLongDeltaV	LongDeltaV
HighestLatDeltaV, SecondLatDeltaV	LatDeltaV
HighestEnergy, SecondEnergy	EnergyDeltaV
HighestBarrierSpeed, SecondBarrierSpeed	BarrierDeltaV

The records extracted from VEHICLECRASH and VEHICLEDEFORMATION are concatenated to form a new table. For records where enough information is provided, a CDC is generated based on the fields DirectionOfForce, DeformationLocation, LongLatLocation, VertLatLocation, DamageDistribution, and DamageExtent, using the CDCLOOKUP table.

Preparing the vehicle crush profiles

Averaged vehicle crush profiles are stored in the table DAMAGELOCATOR. To facilitate the merge between the crush and results tables, the CDCID field is renamed to the DeformID. Some additional formatting is made to ensure that the crush values are consistent with the WinSMASH format. This includes setting all unknown values for the C fields to -10. In WinSMASH, all negative crush values are entered as zero, which allows WinSMASH to use negative crush values as a flag to indicate and unknown or unused crush value.

Any records (usually a small number) with duplicate CaseID, VehicleID, and DeformID values are eliminated from the table as part of this sorting operation. These duplicates typically have a unique ImpactID, but the CDCID that is needed to perform a merge was the same.

Combining the results, crush, and vehicle information

To prepare the final table of WinSMASH inputs, each of the three previously created intermediary tables are merged. Since there are far more records in the vehicle information table than in the others, a large number of extra records are generated by this merge. These records are removed by eliminating all records for which the ΔV is unknown or zero (0). This yields a table containing all of the information needed to define (in WinSMASH) every vehicle involved in every case selected for reconstruction. However, the vehicles still need to be matched up into striking-struck pairs.

In order to pair off the vehicles, this table is sorted by the fields CaseID, EventSequenceID, and VehicleNumber respectively. This changes the organization of the table so that vehicles involved in the same event appear sequentially. The first vehicle for a given EventSequenceID is assigned as WinSMASH vehicle 1. If there is an additional vehicle for that event ID, then it becomes WinSMASH vehicle 2. Both WinSMASH and NASS do not allow for more than two vehicles in an event, so there are never WinSMASH vehicle numbers higher than 2. The WinSMASH vehicle numbers have no correlation to the Oracle vehicle number.

After the WinSMASH IDs are assigned, the table is split into two smaller tables, V1 and V2. V1 contains all of the records that were assigned as WinSMASH vehicle 1, and is left unchanged. V2 contains all of the WinSMASH vehicle 2 records; the names of all the fields are changed so as to be different from those in V1, and V1 and V2 are then merged by CaseID and EventSequenceID. This yields a table containing all the information needed by WinSMASH to define both the striking and struck vehicle for each case to be reconstructed.

The final step in preparing the table is to check each record to verify that it contains enough information to create a WinSMASH run. Records for both vehicles lack results, CDCs, weights, damage lengths, and/or crush profiles are removed from the table. Any null values are filled in with WinSMASH-appropriate placeholders to indicate missing values: Table 18 lists the specific placeholders used for each field.

Table 18. WinSMASH defaults for unknown fields.

Field	Default Value	Field	Default Value
Year	0	Weight	0
Make	"no make"	Damage Length (Smash L)	0
Model	"no model"	Damage Offset (Smash D)	0
Bodystyle	"no bodystyle"	Field L +/- D (D)	0
CDC	(empty string)	C1	-10
PDOF	-1	C2	-10
Heading	-1	C3	-10
Wheelbase	0	C4	-10
Length	0	C5	-10
Width	0	C6	-10

Assignment of Calculation Type

WinSMASH possesses several options for calculating ΔV, referred to as calculation types, which allow a greater degree of flexibility in situations where certain vehicle information is missing. The available calculation types (also called runs):

- Standard Run – Two fully defined vehicles
- Barrier Run – One fully defined vehicle
- CDC Only Run – One fully defined vehicle and one CDC only vehicle
- Missing Run – One fully defined vehicle and one missing vehicle

In addition to each of the calculation types, there are several different types of vehicles. Each vehicle is assigned a vehicle type based on the amount of information known, and the vehicle types can limit the range of calculation types that can be employed. The three vehicle types:

- Fully Defined Vehicle – Both the CDC and crush profile are known
- CDC Only Vehicle – A CDC is known and used to estimate the crush profile
- Missing Vehicle – Neither the CDC or the crush profile are known

The calculation type for WinSMASH runs are not directly collected in Oracle (see the next section) so they must be inferred from PickDeltaVBasis. This field is not stored because many cases can be run as several different calculation types. Thus, it is necessary for AutoSMASH to determine which calculation type (Standard, Missing, Barrier, or CDC Only) was used to reconstruct each case using other information.

The field PickDeltaVBasis, for both vehicle 1 and vehicle 2, is used to determine the calculation type (see Table 19). PickDeltaVBasis values of 1 or 2 correspond to Standard or Barrier cases depending on the number of vehicles in the event. A value of 3 represents a Missing case and 4 represents a CDC Only case. If both vehicles have the same value of 3 or 4, the calculation type is ambiguous because wither vehicle 1 or vehicle 2 could be the missing/CDC Only vehicle.

In situations where the CalcType is ambiguous, an attempt is made to determine which vehicle is Missing or CDC Only and which is Fully Defined by examining the crush profiles; the Missing/CDC Only vehicle often lacks a defined crush profile, while the Fully Defined vehicle should clearly have one. If the calculation type still remains ambiguous, the calculation type is set according to the mapping in Table 19.

Table 19. Calculation type is assigned based on the value of PickDeltaVBasis for each vehicle. Asterisks (*) indicate a calculation type defined by default to an ambiguous combination of PickDeltaVBasis values.

PickDeltaVBasis	PickDeltaVBasis2	CalcType
1	1	Standard
1	2	Standard
1	3	Missing2
1	4	CDC2
2	1	Standard
2	2	Standard
2	3	Missing2
2	4	CDC2
3	1	Missing1
3	2	Missing1
3	3	*Missing2
3	4	Missing1
4	1	CDC1
4	2	CDC1
4	3	Missing2
4	4	*CDC2
-	1	Barrier1
-	2	Barrier1
-	3	Barrier1
-	4	Barrier1
1	-	Barrier2
2	-	Barrier2
3	-	Barrier2
4	-	Barrier2

WinSMASH Tables are Unsuitable for Obtaining Case Reconstruction Inputs

The Oracle database contains a set of "WinSMASH" tables that are meant to collect all of the information needed to redo the NASS cases. There are several WinSMASH tables, but the tables pertinent to AutoSMASH are:

- WSGENERAL
- WSVEHICLE
- WSSPECIFICATIONS
- WSDAMAGE
- WSRESULTS

Unsurprisingly, data in the WinSMASH tables is conveniently laid out in roughly the same way as in the data structure used for WinSMASH input. Additionally, the WinSMASH tables contain certain pieces of information (Calculation type, stiffness, CDC) that are completely missing from or are in different formats in the other tables of the Oracle database. Unfortunately, there are two major problems that prevent the use of this seemingly ideal source for WinSMASH reconstruction parameters.

First, a large number of the runs in the WinSMASH tables are missing damage length (Smash L) values. Every WinSMASH reconstruction has at least one vehicle with a defined damage length, so this parameter is quite important. Fortunately, it is also recorded (much more reliably) in the DAMAGELOCATOR table.

Second, the results in WSRESULTS often do not match the results stored in the VEHICLEDEFORMATION and VEHICLECRASH tables. This is most likely due to manual editing of the results stored in the VEHICLE tables after saving the WinSMASH results to the WinSMASH tables. In such cases, the final version of a run as recorded in the Oracle tables from which ΔV is typically retrieved is different from the version recorded in the WinSMASH tables.

Appendix D - WinSMASH Data Tables

For known occupants with unknown weights, use the occupant's age in the table below to determine the appropriate weight to add.

Table 20. Table of weights to be used for known occupants with unknown weight.

Age (months)	0-2	3-5	6-8	9-11
Weight (Male)	5.4	7.1	8.5	9.8
Weight (Female)	4.9	6.9	8.0	9.1

Age (years)	1	2	3	4	5	6	7	8	9	10	11	12	13	14
Weight (Male)	11.1	13.7	16	18.2	20.7	22.7	25.7	30.4	34.1	36.1	42.1	46.3	53	61
Weight (Female)	10.6	12.9	15	17.2	19.2	21.5	24.7	29.1	34.1	38.3	44.9	49.7	55.5	56.3

Age (years)	15	16	17	18	19	20-29	30-39	40-49	50-59	60-69	70-79	>=80
Weight (Male)	64	69.4	72.9	70.6	73.8	80.2	83.1	85.7	86.4	86.4	81.2	74.7
Weight (Female)	57.6	59.1	59.3	60.9	64.1	67.7	68.8	72.5	73.4	73.5	69.6	62.4

Note – all weights are in kilograms based on 50th percentile for each age group.

Reference:

McDowell, M. A., Fryar, C. D., Hirsch, R., & Ogden, C. L. (2005, July 7). Anthropometric Reference Data for Children and Adults: U.S. Population, 1999-2002. *Advanced Data from Vital and Health Statistics*, Number 361. Atlanta, GA: Centers for Disease Control and Prevention, National Center for Health Statistics.

Appendix E – WinSMASH Research Studies

The research team has conducted several studies on the reconstruction techniques used in WinSMASH, improvements to WinSMASH and the resulting improvements in reconstruction accuracy. The results of these studies have been published in the following technical papers:

Hampton, C. E., & Gabler, H. C. (2010). "Evaluation of the accuracy of NASS/CDS delta-V estimates from the enhanced WinSMASH algorithm," *Annals of Advances in Automotive Medicine,* v.54, pp. 241-252.

Hampton, C. E., & Gabler, H. C. (2009). "NASS/CDS delta-V estimates: The influence of enhancements to the WinSMASH crash reconstruction code," *Annals of Advances in Automotive Medicine,* v.53, pp. 91-102.

Hampton, C. E., & Gabler, H. C. (2009)."Influence of the missing vehicle and CDC-only delta-V reconstruction algorithms for prediction of occupant injury risk," *Biomedical Sciences Instrumentation,* 45: pp. 238-43.

Niehoff, P., & Gabler, H. C. (2006, October). "The accuracy of WinSMASH delta-V estimates: The influence of vehicle type, stiffness, and impact mode," *Annual Proceedings, Association for the Advancement of Automotive Medicine,* 2006, pp. 73-89.

DOT HS 811 546
July 2012

U.S. Department
of Transportation
**National Highway
Traffic Safety
Administration**

8238-070912-v5